Geometry *of* Time-Spaces

Non-commutative Algebraic Geometry,
Applied to Quantum Theory

T0350153

Olav Arnfinn Laudal

University of Oslo, Norway

World Scientific

NEW JERSEY • LONDON • SINGAPORE • BEIJING • SHANGHAI • HONG KONG • TAIPEI • CHENNAI

Geometry *of* Time-Spaces

Non-commutative Algebraic Geometry, Applied to Quantum Theory

Published by

World Scientific Publishing Co. Pte. Ltd.

5 Toh Tuck Link, Singapore 596224

USA office: 27 Warren Street, Suite 401-402, Hackensack, NJ 07601

UK office: 57 Shelton Street, Covent Garden, London WC2H 9HE

British Library Cataloguing-in-Publication Data
A catalogue record for this book is available from the British Library.

GEOMETRY OF TIME-SPACES
Non-commutative Algebraic Geometry, Applied to Quantum Theory

ISBN-13 978-981-4343-34-3
ISBN-10 981-4343-34-X

Printed in Singapore by World Scientific Printers.

This book is dedicated to my grandsons, Even and Amund, and to those few persons in mathematics, that, through the last 18 years, have encouraged this part of my work.

Preface

This book is the result of the author's struggle to understand modern physics. It is inspired by my readings of standard physics literature, but is, really, just a study of the mathematical notion of moduli, based upon my version of non-commutative algebraic geometry. Physics enters in the following way: If we want to study a *phenomenon*, P, in the real world, we have, since Galileo Galilei, been used to associate to P a mathematical *object X*, the *mathematical model* of P, assumed to contain all the information we would like to extract from P. The isomorphism classes, $[X]$, of such objects X, form a *space* $\mathbf{M}$, the *moduli space of the objects X*, on which we may put different *structures*. The assumptions made, makes it reasonable to look for a *dynamical structure*, which to every point $x = [X] \in \mathbf{M}$, *prepared* in some well defined manner, creates a (directed) curve in $\mathbf{M}$, through x, modeling the future of the phenomenon P. Whenever this works, *time* seems to be a kind of metric, on the space, $\mathbf{M}$, measuring all changes in P. It turns out that *non-commutative algebraic geometry*, in my tapping, furnishes, in many cases, the necessary techniques to construct, both the moduli space $\mathbf{M}$, and a universal dynamical structure, $Ph^{\infty}(\mathbf{M})$, from which we may deduce both time and dynamics for non-trivial models in physics. See the introduction for a thorough explanation of the terms used here. The fact that the introduction of a *non-commutative deformation theory*, the basic ingredient in my version of non-commutative algebraic geometry, might lead to a better understanding of the part of modern physics that I had never understood before, occurred to me during a memorable stay at the University of Catania, Italy in 1992. To check this out, has since then been my main interest, and hobby.

Fayence June 2010.

Olav Arnfinn Laudal

Contents

Chapter 1

Introduction

1.1 Philosophy

In a first paper on this subject, see [20], we sketched a *toy model* in physics, where the space-time of classical physics became a section of a universal fiber space $\tilde{E}$, defined on the moduli space, $\mathbf{H} := \mathbf{Hilb}^{(2)}(\mathbf{E}^3)$, of the physical systems we chose to consider, in this case the systems composed of an observer and an observed, both sitting in Euclidean 3-space, $\mathbf{E}^3$. This moduli space is easily computed, and has the form $\mathbf{H} = \underline{\tilde{H}}/Z_2$, where $H = k[t_1, ..., t_6], k = \mathbf{R}$ and $\underline{H} := Spec(H)$ is the space of all ordered pairs of points in $\mathbf{E}^3$, $\underline{\tilde{H}}$ is the blow-up of the diagonal, and Z_2 is the obvious group-action. The space $\mathbf{H}$, and by extension, $\underline{H}$ and $\underline{\tilde{H}}$, was called the *time-space* of the model.

Measurable time, in this mathematical model, turned out to be a metric ρ on the time-space, measuring all possible infinitesimal changes of *the state* of the objects in the family we are studying. A relative velocity is now an oriented line in the tangent space of a point of $\underline{\tilde{H}}$. Thus the space of velocities is compact.

This lead to a *physics* where there are no infinite velocities, and where the principle of relativity comes for free. The Galilean group, acts on $\mathbf{E}^3$, and therefore on $\underline{\tilde{H}}$. The Abelian Lie-algebra of translations defines a 3-dimensional distribution, $\tilde{\Delta}$ in the tangent bundle of $\underline{\tilde{H}}$, corresponding to 0-velocities. Given a metric on $\underline{\tilde{H}}$, we define the distribution $\tilde{c}$, corresponding to light-velocities, as the normal space of $\tilde{\Delta}$. We explain how the classical *space-time* can be thought of as the universal space restricted to a subspace $\underline{\tilde{S}}(l)$of $\underline{\tilde{H}}$, defined by a fixed line $l \subset \mathbf{E}^3$. In chapter 4, under the section *Time-Space and Space-Times*, we shall also show how the generator $\tau \in Z_2$, above, is linked to the operators C, P, T in classical physics, such that

1

$\tau^2 = \tau PT = id$. Moreover, we observe that the three fundamental gauge groups of current quantum theory $U(1)$, $SU(2)$ and $SU(3)$ are part of the structure of the fiber space,

$$\tilde{E} \longrightarrow \tilde{\underline{H}}.$$

In fact, for any point $\underline{t} = (o, x)$ in $\underline{H}$, outside the diagonal Δ, we may consider the line l in $\mathbf{E}^3$ defined by the pair of points $(o, x) \in \mathbf{E}^3 \times \mathbf{E}^3$. We may also consider the action of $U(1)$ on the normal plane $B_o(l)$, of this line, oriented by the normal (o, x), and on the same plane $B_x(l)$, oriented by the normal (x, o). Using parallel transport in $\mathbf{E}^3$, we find an isomorphisms of bundles,

$$P_{o,x} : B_o \to B_x, P : B_o \oplus B_x \to B_o \oplus B_x,$$

the *partition isomorphism*. Using P we may write, (v, v) for $(v, P_{o,x}(v) = P((v, 0))$. We have also seen, in loc.cit., that the line l defines a unique sub scheme $\underline{H}(l) \subset \underline{H}$. The corresponding tangent space at (o, x), is called $A_{(o,x)}$. Together this define a decomposition of the tangent space of $\underline{H}$,

$$T_{\underline{H}} = B_o \oplus B_x \oplus A_{(o,x)}.$$

If $\underline{t} = (o, o) \in \underline{\Delta}$, and if we consider a point $\underline{o}'$ in the exceptional fiber E_o of $\tilde{\underline{H}}$ we find that the tangent bundle decomposes into,

$$T_{\tilde{\underline{H}},o'} = C_{o'} \oplus A_{o'} \oplus \tilde{\Delta},$$

where $C_{o'}$ is the tangent space of E_o, $A_{o'}$ is the light velocity defining o' and $\tilde{\Delta}$ is the 0-velocities. Both B_o and B_x as well as the bundle $C_{(o,x)} := \{(\psi, -\psi) \in B_o \oplus B_x\}$, become complex line bundles on $\underline{H} - \underline{\Delta}$. $C_{(o,x)}$ extends to all of $\tilde{\underline{H}}$, and its restriction to E_o coincides with the tangent bundle. Tensorising with $C_{(o,x)}$, we complexify all bundles. In particular we find complex 2-bundles $\mathbf{C}B_o$ and $\mathbf{C}B_x$, on $\underline{H} - \underline{\Delta}$, and we obtain a canonical decomposition of the complexified tangent bundle. Any real metric on $\underline{H}$ will decompose the tangent space into the light-velocities $\tilde{c}$ and the 0-velocities, $\tilde{\Delta}$, and obviously,

$$T_{\underline{H}} = \tilde{c} \oplus \tilde{\Delta}, \ \mathbf{C}T_{\underline{H}} = \mathbf{C}\tilde{c} \oplus \mathbf{C}\tilde{\Delta}.$$

This decomposition can also be extended to the complexified tangent bundle of $\tilde{\underline{H}}$. Clearly, $U(1)$ acts on $T_{\underline{H}}$, and $SU(2)$ and $SU(3)$ acts naturally on $\mathbf{C}B_o \oplus \mathbf{C}B_x$ and $\mathbf{C}\tilde{\Delta}$ respectively. Moreover $SU(2)$ acts on $\mathbf{C}C_{o'}$, in such a way that their actions should be *physically* irrelevant. $U(1)$, $SU(2)$, $SU(3)$ are our elementary *gauge groups*.

The above example should be considered as the most elementary one, seen from the point of view of present day physics. In fact, whenever we try to make sense of something happening in nature, we consider ourselves as observing something else, i.e. we are working with an observer and an observed, in some sort of ambient space, and the most intuitively acceptable such space, today, is obviously the 3-dimensional Euclidean space.

However, the general philosophy behind this should be the following. If we want to study a natural phenomenon, called **P**, we would, in the present scientific situation, have to be able to describe **P** in some mathematical terms, say as a mathematical object, X, depending upon some parameters, in such a way that the changing aspects of **P** would correspond to altered parameter-values for X. X would be a *model for* **P** if, moreover, X with any choice of parameter-values, would correspond to some, possibly occurring, aspect of **P**.

Two mathematical objects X(1), and X(2), corresponding to the same aspect of **P**, would be called equivalent, and the set, **M**, of equivalence classes of these objects should be called the *moduli space* of the models, X. The study of the natural phenomenon **P**, would then be equivalent to the study of the *structure* of **M**. In particular, the notion of *time* would, in agreement with Aristotle and St. Augustin, see [20], be a *metric* on this space.

With this philosophy, and this *toy*-model in mind we embarked on the study of moduli spaces of representations (modules) of associative algebras in general, see Chapter 3.

Introducing the notion of *dynamical structure*, on the space, **M**, as we shall in (4.1), via the construction of *Phase Spaces*, see Chapter 2, we then have a complete theoretical framework for studying the phenomenon **P**, together with its dynamics.

1.2   Phase Spaces, and the Dirac Derivation

For any associative k-algebra A we have, in [20], and Chapter 2, defined a *phase space* $Ph(A)$, i.e. a universal pair of a morphism $\iota : A \to Ph(A)$, and an ι- derivation, $d : A \to Ph(A)$, such that for any morphism of algebras, $A \to R$, any derivation of A into R decomposes into d followed by an A- homomorphism $Ph(A) \to R$, see [20], and [21]. These associative k-algebras are either trivial or non-commutative. They will give us a natural framework for quantization in physics. Iterating this construction we obtain a limit morphism $\iota^n : Ph^n(A) \to Ph^\infty(A)$ with image $Ph^{(n)}(A)$, and a universal derivation $\delta \in Der_k(Ph^\infty(A), Ph^\infty(A))$, the *Dirac derivation*. This Dirac derivation will, as we shall see, create the dynamics in our different geometries, on which we shall build our theory. For details, see Chapter 2. Notice that the notion of *superspace* is easily deduced from the the Ph-construction. An affine superspace corresponds to a quotient of some $Ph(A)$, where A is the affine k-algebra of some scheme.

1.3 Non-commutative Algebraic Geometry, and Moduli of Simple Modules

The basic notions of affine non-commutative algebraic geometry related to a (not necessarily commutative) associative k-algebra, for k an arbitrary field, have been treated in several texts, see [16], [17], [18], [19]. Given a finitely generated algebra A, we prove the existence of a non-commutative scheme-structure on the set of isomorphism classes of simple finite dimensional representations, i.e. right modules, $Simp_{<\infty}(A)$. We show in [18], and [19], that any *geometric k-algebra* A, see Chapter 3, may be recovered from the (non-commutative) structure of $Simp_{<\infty}(A)$, and that there is an underlying quasi-affine (commutative) scheme-structure on each component $Simp_n(A) \subset Simp_{<\infty}(A)$, parametrizing the simple representations of dimension n, see also [24], [25]. In fact, we have shown that there is a commutative algebra $C(n)$ with an open subvariety $U(n) \subseteq Simp_1(C(n))$, an étale covering of $Simp_n(A)$, over which there exists a *universal representation* $\tilde{V} \simeq C(n) \otimes_k V$, a vector bundle of rank n defined on $Simp_1(C(n))$, and a *versal family*, i.e. a morphism of algebras,

$$\tilde{\rho} : A \longrightarrow End_{C(n)}(\tilde{V}) \to End_{U(n)}(\tilde{V}),$$

inducing all isoclasses of simple n-dimensional A-modules.

Suppose, in line with our *Philosophy* that we have uncovered the moduli space of the mathematical models of our subject, and that A is the *affine k-algebra* of this space, assumed to contain all the parameters of our interest, then the above construction furnishes the *Geometric landscape* on which our *Quantum Theory* will be based.

Obviously, $End_{C(n)}(\tilde{V}) \simeq M_n(C(n))$, and we shall use this isomorphism without further warning.

1.4 Dynamical Structures

We have, above, introduced moduli spaces, both for our mathematical objects, modeling the physical realities, and for the dynamical variables of interest to us. Now we have to put these things together to create dynamics in our geometry.

A *dynamical structure*, see Definition (3.1), defined for a *space*, or any associative k-algebra A, is now an ideal $(\sigma) \subset Ph^\infty(A)$, stable under the Dirac derivation, and the quotient algebra $\mathbf{A}(\sigma) := Ph^\infty(A)/(\sigma)$, will be called a *dynamical system*.

These associative, but usually highly non-commutative, k-algebras are the models for the basic *affine algebras* creating the geometric framework of our theory.

As an example, assume that A is generated by the *space-coordinate functions*, $\{t_i\}_{i=1}^d$ of some *configuration space*, and consider a system of equations,

$$\delta^n t_p := d^n t_p = \Gamma^p(\underline{t}_i, \underline{d}t_j, \underline{d}^2 t_k, .., \underline{d}^{n-1} t_l), \quad p = 1, 2, ..., d.$$

Let $(\sigma) := (\delta^n t_p - \Gamma^p)$ be the two-sided δ-stable ideal of $Ph^\infty(A)$, generated by the equations above, then (σ) will be called a *dynamical structure or a force law, of order n*, and the k-algebra,

$$\mathbf{A}(\sigma) := Ph^\infty(A)/(\sigma),$$

will be referred to as *a dynamical system of order n*.

Producing dynamical systems of interest to physics, is now a major problem. One way is to introduce the notion of *Lagrangian*, i.e. any element $L \in Ph^\infty(A)$, and consider the *Lagrange equation*,

$$\delta(L) = 0.$$

Any δ-stable ideal $(\sigma) \subset Ph^\infty(A)$, for which $\delta(L) = 0 \ (mod(\sigma))$, will be called a *solution of the Lagrange equation*. This is the non-commutative way of taking care of the *parsimony principles* of Maupertuis and Fermat in physics.

In the commutative case, the Dirac derivation of dynamical systems of order 2 will have the form,

$$\delta = \sum_i dt_i \frac{\partial}{\partial t_i} + d^2 t_i \frac{\partial}{\partial dt_i},$$

Whenever A is commutative and smooth, we may consider classical Lagrangians, like, $L = 1/2 \sum_{i,j} g_{i,j} dt_i d_j \in Ph(A)$, a non degenerate metric, expressed in some regular coordinate system $\{t_i\}$. Then the Lagrange equations, produces a dynamical structure of order 2,

$$d^2 t_i = -\sum_{j,k} \Gamma^i_{j,k} dt_j dt_k,$$

where Γ is given by the Levi-Civita connection.

One may also, for a general Lagrangian, $L \in Ph^2(A)$ impose δ as the time, and use the Euler-Lagrange equations, and obtain force laws, see the discussion later in this introduction, and in the section *(4.5) General Quantum Fields, Lagrangians and Actions*.

By definition, δ induces a derivation $\delta_\sigma \in Der_k(\mathbf{A}(\sigma), \mathbf{A}(\sigma))$, also called the *Dirac derivation*, and usually just denoted δ.

For different Lagrangians, we may obtain different Dirac derivations on the same k-algebra $\mathbf{A}(\sigma)$, and therefore, as we shall see, different dynamics of the universal families of the different components of $Simp_n(\mathbf{A}(\sigma))$, $n \geq 1$, i.e. for the *particles* of the system.

1.5 Quantum Fields and Dynamics

Any family of components of $Simp(\mathbf{A}(\sigma))$, with its versal family $\tilde{V}$, will, in the sequel, be called a *family of particles*. A section ϕ of the bundle $\tilde{V}$, is now a function on the moduli space $Simp(A)$, not just a function on the *configuration space*, $Simp_1(A)$, nor $Simp_1(\mathbf{A}(\sigma))$. The value $\phi(v) \in \tilde{V}(v)$ of ϕ, at some point $v \in Simp_n(A)$, will be called a *state* of the particle, at the *event* v.

$End_{C(n)}(\tilde{V})$ induces also a bundle, of *operators*, on the étale covering $U(n)$ of $Simp_n(\mathbf{A}(\sigma))$. A section, ψ of this bundle will be called a *quantum field*. In particular, any element $a \in \mathbf{A}(\sigma)$ will, via the versal family map, $\tilde{\rho}$, define a quantum field, and the set of quantum fields form a k-algebra.

Physicists will tend to be uncomfortable with this use of their language. A classical quantum field for any traditional physicist is, usually, a *function* ψ, defined on some *configuration space*, (which is not our $Simp_n(\mathbf{A}(\sigma))$), with values in the polynomial algebra generated by certain *creation* and *annihilation*-operators in a *Fock-space*.

As we shall see, this interpretation may be viewed as a special case of our general set-up. But first we have to introduce Planck's constant(s) and Fock-space. Then in the section *(4.6) Grand picture, Bosons, Fermions, and Supersymmetry*, this will be explained. There we shall also focus on the notion of locality of interaction, see [11] p. 104, where Cohen-Tannoudji gives a very readable explanation of this strange non-quantum phenomenon in the classical theory, see also [30], the historical introduction.

Notice also that in physics books, the Greek letter ψ is usually used for states, i.e. sections of $\tilde{V}$, or in singular cases, see below, for elements of the Hilbert space, on which their observables act, but it is also commonly used for quantum fields. Above we have a situation where we have chosen to call the quantum fields ψ, reserving ϕ for the states. This is also our language in the section *(4.6) Grand picture, Bosons, Fermions, and Supersymmetry*. Other places, we may turn this around, to fit better with the comparable notation used in physics.

Let $v \in Simp_n(\mathbf{A}(\sigma))$ correspond to the right $\mathbf{A}(\sigma)$-module V, with structure homomorphism $\rho_v : \mathbf{A}(\sigma) \to End_k(V)$, then the Dirac derivation δ composed with ρ_v, gives us an element,

$$\delta_v \in Der_k(\mathbf{A}(\sigma), End_k(V)).$$

Recall now that for any k-algebra B, and right B-modules V, W, there is an exact sequence,

$$Hom_B(V, W) \to Hom_k(V, W) \to Der_k(B, Hom_k(V, W)) \to Ext_B^1(V, W) \to 0,$$

where the image of,

$$\eta : Hom_k(V, W) \to Der_k(B, Hom_k(V, W))$$

is the sub-vectorspace of trivial (or inner) derivations.

Modulo the trivial (inner) derivations, δ_v defines a class,

$$\xi(v) \in Ext_{\mathbf{A}(\sigma)}^1(V, V),$$

i.e. a tangent vector to $Simp_n(\mathbf{A}(\sigma))$ at v. The Dirac derivation δ therefore defines a unique one-dimensional distribution in $\Theta_{Simp_n(\mathbf{A}(\sigma))}$, which, once we have fixed a versal family, defines a vector field,

$$\xi \in \Theta_{Simp_n(\mathbf{A}(\sigma))},$$

and, in good cases, a (rational) derivation,

$$\xi \in Der_k(C(n))$$

inducing a derivation,

$$[\delta] \in Der_k(\mathbf{A}(\sigma), End_{C(n)}(\tilde{V})),$$

lifting ξ, and, in the sequel, identified with ξ. By definition of $[\delta]$, there is now a *Hamiltonian operator*

$$Q \in M_n(C(n)),$$

satisfying the following fundamental equation, see Theorem (4.2.1),

$$\delta = [\delta] + [Q, \tilde{\rho}(-)].$$

This equation means that for an element (an observable) $a \in \mathbf{A}(\sigma)$ the element $\delta(a)$ acts on $\tilde{V} \simeq C(n)^n$ as $[\delta](a) = \xi(\tilde{\rho}_V(a))$ plus the Lie-bracket $[Q, \tilde{\rho}_V(a)]$.

Notice that any right $\mathbf{A}(\sigma)$-module V is also a $Ph^\infty(A)$-module, and therefore corresponds to a family of $Ph^n(A)$-module-structures on V, for $n \geq 1$, i.e. to an A-module $V_0 := V$, an element $\xi_0 \in Ext^1_A(V,V)$, i.e. a tangent of the deformation functor of $V_0 := V$, as A-module, an element $\xi_1 \in Ext^1_{Ph(A)}(V,V)$, i.e. a tangent of the deformation functor of $V_1 := V$ as $Ph(A)$-module, an element $\xi_2 \in Ext^1_{Ph^2(A)}(V,V)$, i.e. a tangent of the deformation functor of $V_2 := V$ as $Ph^2(A)$-module, etc. All this is just V, considered as an A-module, together with a sequence $\{\xi_n\}, 0 \leq n$, of a tangent, or a *momentum*, ξ_0, an acceleration vector, ξ_1, and any number of higher order *momenta* ξ_n. Thus, specifying a point $v \in Simp_n(A(\sigma))$ implies specifying a *formal curve* through v_0, the base-point, of the *miniversal deformation space* of the A-module V.

Knowing the dynamical structure, (σ), and the state of our *object V* at a *time* τ_0, i.e. knowing the structure of our *representation V* of the algebra $\mathbf{A}(\sigma)$, at that time (which is a problem that we shall return to), the above makes it reasonable to believe that we, from this, may deduce the state of V at any *later* time τ_1. This assumption, on which all of science is based, is taken for granted in most textbooks in modern physics. This paper is, in fact, an attempt to give this basic assumption a reasonable basis. The mystery is, of course, why Nature seems to be parsimonious, in the sense of Fermat and Maupertuis, giving us a chance of guessing dynamical structures.

The *dynamics of the system* is now given in terms of the Dirac vector-field $[\delta]$, generating the vector field ξ on $Simp_n(\mathbf{A}(\sigma))$. An integral curve γ of ξ is a *solution* of the *equations of motion*. Let γ start at $v_0 \in Simp_n(\mathbf{A}(\sigma))$ and end at $v_1 \in Simp_n(\mathbf{A}(\sigma))$, with length $\tau_1 - \tau_0$. This is only meaningful for ordered fields k, and when we have given a metric (time) on the moduli space $Simp_n(\mathbf{A}(\sigma))$. Assume this is the situation. Then, given a *state*, $\phi(v_0) \in \tilde{V}(v_0) \simeq V_0$, of the *particle* $\tilde{V}$, we prove that there is a *canonical evolution map*, $U(\tau_0, \tau_1)$ transporting $\phi(v_0)$ from time τ_0, i.e. from the point representing V_0, to time τ_1, i.e. corresponding to some point representing V_1, along γ. It is given as,

$$U(\tau_0, \tau_1)(\phi(v_0)) = exp(\int_\gamma Q d\tau)(\phi(v_0)),$$

where $exp(\int_\gamma)$ is the non-commutative version of the classical *action integral*, related to the Dyson series, to be defined later, see the proof of Theorem (4.2.3) and the section *(4.6) Grand picture. Bosons, Fermions, and Supersymmetry*. In case we work with unitary representations, of some sort, we may also deduce analogies to the *S*-matrix, perturbation theory, and so also to Feynman-integrals and diagrams.

1.6 Classical Quantum Theory

Most of the classical models in physics are either essentially commutative, or *singular*, i.e. such that either $Q = 0$, or $[\delta] = 0$. General relativity is an example of the first category, classical Yang-Mills theory is of the second kind. In fact, any theory involving connections are singular, and infinite dimensional. But we shall see that imposing singularity on a theory, sometimes recover the classical infinite dimensional (Hilbert-space-based) model as a limit of the finite dimensional simple representations, corresponding to a dynamic system, see Examples 4.2-4.4, where we treat the *Harmonic Oscillator*.

1.7 Planck's Constants, and Fock Space

This general model allows us also to define a general notion of a *Planck's constant(s)*, $\hbar_l$, as the generator(s) of the *generalized monoid*,

$$\Lambda(\sigma) := \{\lambda \in C(n) | \exists f_\lambda \in \mathbf{A}(\sigma), f_\lambda \neq 0,$$
$$[Q, \tilde{\rho}(\delta(f_\lambda))] = \tilde{\rho}(\delta(f_\lambda)) - [\delta](\tilde{\rho}(f_\lambda)) = \lambda \tilde{\rho}(f_\lambda)\}$$

which has the property that $\lambda, \lambda' \in \Lambda(\sigma), f_\lambda f_{\lambda'} \neq 0$ implies $\lambda + \lambda' \in \Lambda(\sigma)$. From this definition we may construct a general notion of *Fock algebra*, or *Fock space*, and a representation, both named **F**, on this space. **F** is the sub-k-algebra of $End_{C(n)}(\tilde{V}))$ generated by $\{a_+^l := f_{\hbar_l}, a_-^l := f_{-\hbar_l}\}$, see Examples (4.3) and (4.5) for a rather complete discussion of the one-dimensional harmonic oscillator in all ranks, and of the quartic anharmonic oscillator in rank 2 and 3. Notice that this is just a natural generalization of standard work on classification of representations of (semi-simple) Lie algebras, see the discussion of *fundamental particles* in the Example (4.14).

When A is the coordinate k-algebra of a moduli space, we should also consider the family of Lie algebras of *essential* automorphisms of the objects classified by $Simp(A(\sigma))$, and apply invariant theory, like in [18], to obtain a general form for Yang-Mills theory, see [33] and [22], for the case of plane curve singularities. This would offer us a general model for the notions of *gauge particles* and *gauge fields*, coupling with ordinary *particles* via representations onto corresponding simple modules.

1.8 General Quantum Fields, Lagrangians and Actions

Perfectly parallel with this theory of simple finite dimensional representations, we might have considered, for given algebras A, and B, the space of algebra homomorphisms,

$$\phi : A \to B.$$

In the commutative, classical case, when A is generated by $t_1, ..., t_r$, and B is the affine algebra of a *configuration space* generated by $x_1, ..., x_s$, ϕ is determined by the images $\phi_i := \tilde{\phi}(t_i)$, and ϕ or $\{\phi_i\}$ is called a *classical field*. Any such *field*, ϕ induces a unique commutative diagram of algebras,

$$
\begin{array}{ccc}
A & \xrightarrow{\phi} & B \\
\downarrow & & \downarrow \\
Ph(A) & \xrightarrow{Ph\phi} & Ph(B).
\end{array}
$$

Given dynamical structures, (say of order two), σ and μ, defined on A, respectively B, we construct a vector field $[\delta]$ on the *space*, $\mathbf{F}(A(\sigma), B(\mu))$, of fields, $\phi : A(\sigma) \to B(\mu)$. The singularities of $[\delta]$ defines a subset,

$$\mathbf{M} := \mathbf{M}(A(\sigma), B(\mu)) \subset \mathbf{F}(A(\sigma), B(\mu)) =: \mathbf{F}.$$

There are natural equations of *motion*, analogous to those we have seen above, see (3.2). Notice that a field $\phi \in \mathbf{M}$ is said to be *on shell*, those of $\mathbf{F} - \mathbf{M}$ are *off shell*. We shall explore the structure of $\mathbf{M}$ in some simple cases.

The actual choice of dynamical structures $(\sigma), (\mu)$, for the particular physical set-up, is, of course, not obvious. They may be defined in terms of *force laws*, but, in general, force laws do not pop up naturally. Instead, physicists are used to insist on the Lagrangian, an element $L \in Ph(A)$, as a main player in this game. The *Lagrangian density*, $\mathbf{L}$ should then be considered an element of the versal family of the iso-clases of $\mathbf{F}(A, B)$. In fact, assuming that this space has a local affine algebraic geometric structure, parametrized by some ring $\mathbf{C}$, we may consider the versal family as a homomorphisms of k-algebras,

$$\tilde{\phi} : Ph(A) \to \mathbf{C} \otimes_k Ph(B),$$

and put $\mathbf{L} := \tilde{\phi}(L)$. Classically one picks a (natural) representation, corresponding to a derivation of B,

$$\rho : Ph(B) \to B,$$

and put, $\mathbf{L} := \rho(\mathbf{L})$. One considers the Lagrangian density as a function in $\phi_i, \phi_{i,j} := \frac{\partial \phi_i}{\partial x_j}$, thus as a function on configuration space $Simp_1(B)$, with coefficients from $\mathbf{C}$. One postulates that there is a functional, or an *action*, which, for every field ϕ, associates a real or complex value,

$$S := S(\mathbf{L}(\phi_i, \phi_{i,j})),$$

usually given in terms of a trace, or as an integral of $\mathbf{L}$ on part of the configuration space, see below. S should be considered as a function on $\mathbf{F} := \mathbf{F}(A, B)$, i.e. as an element of $\mathbf{C}$. The parsimony principles of Fermat and Maupertuis is then applied to this function, and one wants to compute the vector field,

$$\nabla S \in \Theta_{\mathbf{F}},$$

which mimic our $[\delta]$, derived from the Dirac derivations. The equation of motion, i.e. the equations picking out the subspace $\mathbf{M} \subset \mathbf{F}$, is therefore,

$$\nabla S = 0.$$

Here is where classical calculus of variation enters, and where we obtain differential equations for ϕ_i, the *Euler-Lagrange equations of motion*.

Notice now that in an infinite dimensional representation, the *Trace* is an integral on the spectrum. The equation of motion defining $\mathbf{M} \subset \mathbf{F}$, now corresponds to,

$$\delta S := \delta \int \tilde{\rho}(\mathbf{L}) = 0.$$

The calculus of variation produces Euler-Lagrange equations, and so picks out the singularities of ∇S, the replacement for $[\delta]$, without referring to a dynamical structure, or to (uni)versal families. See the Examples (3.7) and

(3.8), where we treat the harmonic oscillator, and where we show that the classical infinite dimensional representation is a limit of finite dimensional simple representations. We also show that the Lagrangian of the harmonic oscillator produces a vector field ∇S on $Simp_2(A(\sigma))$ which is different from the one generated by the Dirac derivation for the dynamical system deduced from the Euler-Lagrange equations for the same Lagrangian. However, the sets of singularities for the two vector fields coincide.

This should never the less be cause for worries, since the world we can test is finite. The infinite dimensional mathematical machinery is obviously just a computational trick.

Another problem with this reliance on the Parsimony Principle via Lagrangians, and the (commutative) Euler-Lagrange equations, is that, unless we may prove that $\nabla S = [\delta]$, for some dynamical structure σ, the philosophically satisfying realization, that a preparation in $A(\sigma)$ actually implies a deterministic future for our objects, disappear, see above.

Otherwise, it is clear that the theory becomes more flexible. It is easy to cook up Lagrangians.

In QFT, when quantizing fields, physicists are, however, usually strangely vague; suddenly they consider functions, $\{\phi_i, \phi_{i,j}\}$, on configuration space, as elements in a k-algebra, introduce commutation relations and start working as if these functions on configuration space were operators. This is, maybe, due to the fact that they do not see the difference between the role of B in the classical case, and the role of $Ph(B)$, in quantum theory.

1.9   Grand Picture. Bosons, Fermions, and Supersymmetry

With this done, we sketch the big picture of QFT that emerges from the above ideas. This is then used as philosophical basis for the treatment of the harmonic oscillator, general relativity, electromagnetism, spin and quarks, which are the subjects of the Examples (4.2) to (4.14).

In particular, we sketch, here and in Chapter 5, how we may treat the problems of *Bosons*, *Fermions*, *Anyons*, and *Super-symmetry*.

1.10   Connections and the Generic Dynamical Structure

Moreover we shall see that, on a space with a non-degenerate metric, there is a unique *generic dynamical structure*, (σ), which produces the most interesting physical models. In fact, any connection on a bundle, induces a representation of $\mathbf{A}(\sigma)$. We shall use this metod to quantize the *Electromagnetic Field*, as well as the *Gravitational Field*, obtaining generalized Maxwell, Dirac and Einstein-type equations, with interesting properties, see Examples (4.1), (4.13) and (4.14). The Levi-Civita connection turns out to be a very particular singular representation for which the Hamiltonian is identified with the *Laplace-Beltrami operator*.

1.11   Clocks and Classical Dynamics

At this point we need to be more interested in how to measure time. We therefore discuss the notion of clocks in this picture, and we propose two rather different models, one called *The Western clock*, modeled on a free particle in dimension 1, i.e. one with $d^2\tau = 0$, and another, called the *Eastern Clock*, modeled on the harmonic oscillator in dimension 1, i.e. one with $d^2\tau = \tau$.

1.12   Time-Space and Space-Times

The application to the case of point-like particles in the $\tilde{\underline{H}}$-model is treated in Example (3.5), mainly as an introduction to the study of the Levi-Civita connection, in our tapping. Coupled with the non-trivial geometry of $\tilde{\underline{H}}$ we see a promising possibility of defining notions like mass and charge, of different colors, related to the structure of $\tilde{\underline{H}}$ along the diagonal $\tilde{\underline{\Delta}}$. A catchy way of expressing this would be that every point in our real world is a *black hole*, outfitted with a density of, at least, mass and charge. Notice that the dimension of $\tilde{\underline{\Delta}}$ is 5, which brings about ideas like those of Kaluza and Klein.

In particular, the definition of mass, and the deduction of Newton's law of gravitation, from the assumption that mass is a property of the geometry of $\tilde{\underline{H}}$, related to the blow up along the diagonal, seems promising. A simple example in this direction leads to a Schwarzschild-type geometry. The corresponding equations of motion reduces to Kepler's laws, see the Example (4.12). As another example, we shall again go back to our toy-model, where the standard Gauge groups, $U(1)$, $SU(2)$, and $SU(3)$ pop up canonically and show that the results above can be used to construct a general geometric theory closely related to general relativity and to quantum theory, generalizing both. See the Examples (4.13), (4.14), where the action of

the natural gauge group, on the canonically decomposed tangent bundle of
$\underline{H}$, as described above, sets up a nice theory for *elementary particles, spin,
isospin, hypercharge*, including *quarks*. Here the notion of non-commutative
invariant space, plays a fundamental role. In particular, notice the possible
models for light and dark matter, or energy, hinted upon in the Examples
(4.13), and (4.14). Notice also that, in this toy model light cannot be described as point-particles. There are no radars available for point-particles,
like in current general relativity. However, the quantized E-M works well to
explain communication with light. Moreover, as one might have expected,
a reasonable model of the process creating the universe as we see it, will
provide a better understanding of what we are modeling. This is the subject
of the next section.

1.13 Cosmology, Big Bang and All That

Our toy-model, i.e. the moduli space, **H**, of two points in the Euclidean
3-space, or its étale covering, $\tilde{\underline{H}}$, turns out to be *created* by the versal deformation of the obvious (non-commutative) singularity in 3-dimensions,
$U := k < x_1, x_2, x_3 > /(x_1, x_2, x_3)^2$. In fact, it is easy to see that the versal
space of the deformation functor of the k-algebra U contains a flat component (a room in the modular suite, see [22]) isomorphic to $\tilde{\underline{H}}$. The modular
stratum (the inner room) is reduced to the base point. This furnishes a
nice model for *The Universe* with easy relations to classical cosmological
models, like those of Friedman-Robertson-Walker, and Einstein-de Sitter.

1.14 Interaction and Non-commutative Algebraic Geome-
     try

In section 1.4, we shall introduce interactions, lifetime, decay and creation
of particles. The inspiration for this final paragraph comes from elementary
physics concerning *Cross-Sections, Resonance*, and *The Cluster Decomposition Principle*, see Weinberg, [30], I, (3.8).

   The possibility of treating interaction between fields in a perfectly geometric way, with the usual metrics and connections replaced with a *non-
commutative metric* is, maybe, the most interesting aspect of the model
presented in this paper.

   The essential point is that, in non-commutative algebraic geometry, say
in the *space* of representations of an algebra B, there is a *tangent space*,

$T(V, W) := Ext^1_B(V, W)$, between any two points, V, W. In particular, if $B = Ph(k[x_1, ..., x_n])$, then any 1-dimensional representation of B is represented as a pair (q, ξ), of a closed point q of $Spec(k[\underline{x}])$, and a tangent ξ at that point. Given two such points, $(q_i, \xi_i), i = 1, 2$, an easy calculation proves that $T((q_1, \xi_1), (q_2, \xi_2))$ is of dimension 1 if $q_1 \neq q_2$, of dimension n if $q_1 = q_2, \xi_1 \neq \xi_2$, and of dimension 2n when $(q_1, \xi_1) = (q_2, \xi_2)$, see Example (1.1), (ii).

Now, just as we may talk about vector-fields, as the assignment of a tangent vector to any point in space, and consider metrics as functions that associate a length to any tangent-vector, we may consider fields of tangents between any two points, and extend the notion of metric to measure the length of such.

If we do, we find very nice models for treating the notion of identical particles, and interaction between fields, see the Examples (5.1), (5.2).

Finally, we shall not resist the temptation to attempt a formalization, in our language, of the notion of Alternative Histories, see [6], p.140, and the paper [7]. The result is another application of noncommutative deformation theory which seems to be a promising tool in mathematical physics.

1.15   Apology

Referring to the historical introduction of Weinberg's, *The Quantum Theory of Fields*, see [30], where he quotes Heisenberg's 1925-Manifesto, I must confess that the present paper is based on the same positivistic philosophy as the one Weinberg rules out.

But then, I am not a physicist, and this paper is a paper on geometry of certain finitely generated non-commutative algebraic schemes, where I have taken the liberty of using my version of the physicist jargon to make the results more palpable.

Even though I see a lot of difficulties in the interpretation of the mathematical notions of my models, in a physics context, I hope that the model I propose may help other mathematicians to gain faith in their jugendtraums; sometime, somehow, to be able to understand some physics.

An attentive reader will also see that, if my *modelist philosophy* about Nature, see above, should be taken seriously, it would reduce the physicists work to define, in a mathematical language, the model of the object she is studying, then with the help of a mathematician work out the moduli space of all such models, define the infinite phase space of this moduli space, guess on a metric to define time, and a corresponding dynamical structure, and give the result to the computer algebra group in Kaiserslautern, and hope for the best.

Chapter 2

Phase Spaces and the Dirac Derivation

2.1 Phase Spaces

Given a k-algebra A, denote by $A/k-\underline{alg}$ the category where the objects are homomorphisms of k-algebras $\kappa : A \to R$, and the morphisms, $\psi : \kappa \to \kappa'$ are commutative diagrams,

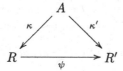

and consider the functor,

$$Der_k(A, -) : A/k - \underline{alg} \longrightarrow \underline{Sets}.$$

It is representable by a k-algebra-morphism,

$$\iota : A \longrightarrow Ph(A),$$

with a *universal family* given by a universal derivation,

$$d : A \longrightarrow Ph(A).$$

Ph (A) is relatively easy to compute. It can be constructed as the non-commutative versal base of the deformation functor of the morphism $\rho : A \to k[\epsilon]$, see [20] and [21].

Clearly we have the identities,

$$d_* : Der_k(A, A) = Mor_A(Ph(A), A),$$

and,

$$d^* : Der_k(A, Ph(A)) = End_A(Ph(A)),$$

17

the last one associating d to the identity endomorphism of Ph. Let now V be a right A-module, with structure morphism $\rho : A \to End_k(V)$. We obtain a universal derivation,

$$c : A \longrightarrow Hom_k(V, V \otimes_A Ph(A)),$$

defined by, $c(a)(v) = v \otimes d(a)$. Using the long exact sequence, see the introduction,

$$0 \to Hom_A(V, V \otimes_A Ph(A)) \to Hom_k(V, V \otimes_A Ph(A))$$
$$\to^\iota Der_k(A, Hom_A(V, V \otimes_A Ph(A))) \to^\kappa Ext^1_A(V, V \otimes_A Ph(A)) \to 0,$$

we obtain the non-commutative *Kodaira-Spencer class,*

$$c(V) := \kappa(c) \in Ext^1_A(V, V \otimes_A Ph(A)),$$

inducing the *Kodaira-Spencer morphism,*

$$g : \Theta_A := Der_k(A, A) \longrightarrow Ext^1_A(V, V),$$

via the identity, $\tilde{\delta}_*$. If $c(V) = 0$, then the exact sequence above proves that there exist a $\nabla \in Hom_k(V, V \otimes_A Ph(A))$ such that $\tilde{\delta} = \iota(\nabla)$. This is just another way of proving that $\tilde{\delta}$ is given by a connection,

$$\nabla : Der_k(A, A) \longrightarrow Hom_k(V, V).$$

As is well known, in the commutative case, the Kodaira-Spencer class gives rise to a *Chern character* by putting,

$$ch^i(V) := 1/i! \; c^i(V) \in Ext^i_A(V, V \otimes_A Ph(A)),$$

and if $c(V) = 0$, the curvature $R(V)$ induces a curvature class,

$$R_\nabla \in H^2(k, A; \Theta_A, End_A(V)).$$

Any $Ph(A)$-module W, given by its structure map,

$$\rho_W : Ph(A) \longrightarrow End_k(W)$$

corresponds bijectively to an induced A-module structure on W, and a derivation $\delta_\rho \in Der_k(A, End_k(W))$, defining an element,

$$[\delta_\rho] \in Ext^1_A(W, W),$$

see the introduction. Fixing this element we find that the set of $Ph(A)$-module structures on the A-module W is in one to one correspondence with,

$$End_k(W)/End_A(W).$$

Conversely, starting with an A-module V and an element $\delta \in Der_k(A, End_k(V))$, we obtain a $Ph(A)$-module V_δ. It is then easy to see that the kernel of the natural map,

$$Ext^1_{Ph(A)}(V_\delta, V_\delta) \to Ext^1_A(V, V),$$

induced by the linear map,

$$Der_k(Ph(A), End_k(V_\delta)) \to Der_k(A, End_k(V))$$

is the quotient,

$$Der_A(Ph(A), End_k(V_\delta))/End_k(V).$$

Remark 2.1. Since $Ext^1_A(V, V)$ is the tangent space of the miniversal deformation space of V as an A-module, see e.g. [18], or the next chapter, we see that the non-commutative space $Ph(A)$ also parametrizes the set of *generalized momenta*, i.e. the set of pairs of a simple module $V \in Simp(A)$, and a tangent vector of $Simp(A)$ at that point.

Example 2.1. (i) Let $A = k[t]$, then obviously, $Ph(A) = k < t, dt >$ and d is given by $d(t) = dt$, such that for $f \in k[t]$, we find $d(f) = J_t(f)$ with the notations of [19], i.e. the non-commutative derivation of f with respect to t. One should also compare this with the non-commutative Taylor formula of loc.cit. If $V \simeq k^2$ is an A-module, defined by the matrix $X \in M_2(k)$, and $\delta \in Der_k(A, End_k(V))$, is defined in terms of the matrix $Y \in M_2(k)$, then the $Ph(A)$-module V_δ is the $k < t, dt >$-module defined by the action of the two matrices $X, Y \in M_2(k)$, and we find

$$e^1_V := dim_k Ext^1_A(V, V) = dim_k End_A(V) = dim_k\{Z \in M_2(k)|\ [X, Z] = 0\}$$
$$e^1_{V_\delta} := dim_k Ext^1_{Ph(A)}(V_\delta, V_\delta) = 8 - 4 + dim\{Z \in M_2(k)|\ [X, Z] = [Y, Z] = 0\}.$$

We have the following inequalities,

$$2 \leq e^1_V \leq 4 \leq e^1_{V_\delta} \leq 8.$$

(ii) Let $A = k^2 \simeq k[x]/(x^2 - r^2)$, $r \in k$, $r \neq 0$, then,

$$Ph(A) = k < x, dx > /((x^2 - r^2), x \cdot dx + dx \cdot x).$$

Notice that $Ph(A)$ just has 2 points, i.e. simple representations, given by,

$$k(r) : x = r, dx = 0, \ k(-r) : x = -r, \ dx = 0.$$

An easy computation shows that,

$$Ext^1_{Ph(A)}(k(\alpha), k(\alpha)) = 0, \alpha = r, -r, \ Ext^1_{Ph(A)}(k(\alpha), k(-\alpha)) = k \cdot \omega,$$

where ω is represented by the derivation given by $\omega(x) = 2r$, $\omega(dx) = t \in k$ where t is the *tension* of this *string* of dimension -1, see end of §2, and end of §3. Notice also that this is an example of the existence of *tangents between different points*, in non-commutative algebraic geometry.

(iii) Now, let $A = k[\underline{x}] := k[x_1, x_2, x_3]$ and consider,

$$Ph(A) = k < x_1, x_2, x_3, dx_1, dx_2, dx_3 > /([x_i, x_j], d([x_i, x_j])).$$

Any rank 1 representation of A is represented by a pair of a closed point q of $Spec(k[\underline{x}])$, and a tangent p at that point. Given two such points, $(q_i, p_i), i = 1, 2$, an easy calculation proves,

$$dim_k Ext^1_{PhA}(k(q_1, p_1), k(q_2, p_2)) = 1, \text{for } q_1 \neq q_2$$
$$dim_k Ext^1_{PhA}(k(q_1, p_1), k(q_2, p_2)) = 3, \text{for } q_1 = q_2, , p_1 \neq p_2$$
$$dim_k Ext^1_{PhA}(k(q_1, p_1), k(q_2, p_2)) = 6, \text{for } (q_1, p_1) = (q_2, p_2)$$

Put $x_j(q_i, p_i) := q_{i,j}$, $dx_j((q_i, p_i) := p_{i,j}$, $\alpha_j = q_{1,j} - q_{2,j}$, $\beta_j = p_{1,j} - p_{2,j}$. See that for any element $\alpha \in Hom_k(k((q_1, p_1)), k((q_2, p_2)))$ we have,

$$x_j \alpha = q_{1,j} \alpha, \quad \alpha x_j = q_{2,j} \alpha, \quad dx_j \alpha = p_{1,j} \alpha, \quad \alpha dx_j = p_{2,j} \alpha,$$

with the obvious identification. Any derivation

$$\delta \in Der_k(PhA, Hom_k(k((q_1, p_1)), k((q_2, p_2))))$$

must satisfy the relations,

$$\delta([x_i, x_j]) = [\delta(x_i), x_j] + [x_i, \delta(x_j)] = 0$$
$$\delta([dx_i, x_j] + [x_i, dx_j]) = [\delta(dx_i), x_j] + [dx_i, \delta(x_j)] + [\delta(x_i), dx_j] + [x_i, \delta(dx_j)] = 0.$$

Using the above left-right action-rules, the result follows from the long exact sequence computing Ext^1_{PhA}. The two families of relations above give us two systems of linear equations.

The first, in the variables $\delta(x_1), \delta(x_2), \delta(x_3), \delta(dx_1), \delta(dx_2), \delta(dx_3)$, with matrix,

$$\begin{pmatrix} -\beta_2 & \beta_1 & 0 & -\alpha_2 & \alpha_1 & 0 \\ -\beta_3 & 0 & \beta_1 & -\alpha_3 & 0 & \alpha_1 \\ -\beta_3 & \beta_2 & 0 & -\alpha_3 & 0 & \alpha_2 \end{pmatrix}$$

and the second, in the variables, $\delta(x_1), \delta(x_2), \delta(x_3)$, with matrix,

$$\begin{pmatrix} -\alpha_2 & \alpha_1 & 0 \\ -\alpha_3 & 0 & \alpha_1 \\ 0 & -\alpha_3 & \alpha_2 \end{pmatrix}.$$

In particular we see that the *trivial* derivation given by,

$$\delta(x_i) = \alpha_i, \quad \delta(dx_j) = \beta_j,$$

satisfies the relations, and the generator of $Ext^1_{PhA}(k(q_1, p_1), k(q_2, p_2))$ is represented by,

$$\delta(x_i) = 0, \quad \delta(dx_j) = \alpha_i.$$

This is, in an obvious sense, the vector $-(q_1, q_2)$, and we notice that this generator is of the type $\delta(d-)$, so it is an acceleration in $Simp_1(k[\underline{x}])$, see the interpretation of this as an *interaction* in Chapter 5. It is not difficult to extend this result from dimension 3 to any dimension n.

(iv) Consider now the space of 2-dimensional representation of $Ph(A)$. It is an easy computation that any such is given by the actions,

$$x_1 = \begin{pmatrix} a_1 & 0 \\ 0 & a_2 \end{pmatrix}, \quad x_2 = \begin{pmatrix} b_1 & 0 \\ 0 & b_2 \end{pmatrix}, \quad x_3 = \begin{pmatrix} c_1 & 0 \\ 0 & c_2 \end{pmatrix},$$

and,

$$dx_1 = \begin{pmatrix} \alpha_{1,1} & (a_1 - a_2) \\ (a_2 - a_1) & \alpha_{2,2} \end{pmatrix},$$

$$dx_2 = \begin{pmatrix} \beta_{1,1} & (b_1 - b_2) \\ (b_2 - b_1) & \beta_{2,2} \end{pmatrix},$$

$$dx_3 = \begin{pmatrix} \gamma_{1,1} & (c_1 - c_2) \\ (c_2 - c_1) & \gamma_{2,2} \end{pmatrix}.$$

The *angular momentum* is now given by,

$$L_{1,2} := x_1 dx_2 - x_2 dx_1 = \begin{pmatrix} (a_1\beta_{1,1} - b_1\alpha_{1,1}) & (a_2 b_1 - a_1 b_2) \\ (a_1 b_2 - a_2 b_1) & (a_2\beta_{2,2} - b_2\alpha_{2,2}) \end{pmatrix},$$

etc. And *the isospin*, see (3.18) and (3.19), has the form,

$$I_1 := [x_1, dx_1] = \begin{pmatrix} 0 & (a_1 - a_2)^2 \\ (a_2 - a_1)^2 & 0 \end{pmatrix},$$

etc.

(v) Let $A = M_2(k)$, and let us compute $Ph(A)$. Clearly the existence of the canonical homomorphism, $i : M_2(k) \to Ph(M_2(k))$ shows that $Ph(M_2(k))$ must be a matrix ring, generated, as an algebra, over $M_2(k)$ by $d\epsilon_{i,j}$, $i, j = 1, 2$, where $\epsilon_{i,j}$ is the elementary matrix. A little computation shows that we have the following relations,

$$d\epsilon_{1,1} = \begin{pmatrix} 0 & (d\epsilon_{1,1})_{1,2} = -(d\epsilon_{2,2})_{1,2} \\ (d\epsilon_{1,1})_{2,1} = -(d\epsilon_{2,2})_{2,1} & 0 \end{pmatrix}$$

$$d\epsilon_{2,2} = \begin{pmatrix} 0 & (d\epsilon_{2,2})_{1,2} = -(d\epsilon_{1,1})_{1,2} \\ (d\epsilon_{2,2})_{2,1} = -(d\epsilon_{1,1})_{2,1} & 0 \end{pmatrix}$$

$$d\epsilon_{1,2} = \begin{pmatrix} \epsilon_{1,2}(d\epsilon_{2,2})_{2,1} & (d\epsilon_{1,2})_{1,2} = -(d\epsilon_{2,1})_{2,1} \\ 0 & -(d\epsilon_{2,2})_{2,1}\epsilon_{1,2} \end{pmatrix}$$

$$d\epsilon_{2,1} = \begin{pmatrix} (d\epsilon_{2,2})_{1,2}\epsilon_{2,1} & 0 \\ (d\epsilon_{2,1})_{2,1} = -(d\epsilon_{1,2})_{1,2} & \epsilon_{2,1}(d\epsilon_{1,1})_{1,2} \end{pmatrix}$$

From this follows that any section, $\rho : Ph(M_2(k)) \to M_2(k)$, of $i : M_2(k) \to Ph(M_2(k))$, is given in terms of an element $\phi \in M_2(k)$, such that $\rho(da) = [\phi, a]$.

2.2 The Dirac Derivation

The phase-space construction may, of course, be iterated. Given the k-algebra A we may form the sequence, $\{Ph^n(A)\}_{1 \leq n}$, defined inductively by

$$Ph^0(A) = A, \ Ph^1(A) = Ph(A), ..., Ph^{n+1}(A) := Ph(Ph^n(A)).$$

Let $i_0^n : Ph^n(A) \to Ph^{n+1}(A)$ be the canonical imbedding, and let $d_n : Ph^n(A) \to Ph^{n+1}(A)$ be the corresponding derivation. Since the composition of i_0^n and the derivation d_{n+1} is a derivation $Ph^n(A) \to Ph^{n+2}(A)$, there exist by universality a homomorphism $i_1^{n+1} : Ph^{n+1}(A) \to Ph^{n+2}(A)$, such that,

$$d_n \circ i_1^{n+1} = i_0^n \circ d_{n+1}.$$

Notice that we here compose functions and functors from left to right. Clearly we may continue this process constructing new homomorphisms,

$$\{i_j^n : Ph^n(A) \to Ph^{n+1}(A)\}_{0 \leq j \leq n},$$

with the property,

$$d_n \circ i_{j+1}^{n+1} = i_j^n \circ d_{n+1}.$$

It is easy to see, [21], that,

$$i_p^n i_q^{n+1} = i_{q-1}^n i_p^{n+1}, \ p < q$$
$$i_p^n i_p^{n+1} = i_p^n i_{p+1}^{n+1}$$
$$i_p^n i_q^{n+1} = i_q^n i_{p+1}^{n+1}, \ q < p,$$

i.e. the Ph^*A is a semi-cosimplicial algebra. The system of k-algebras and homomorphisms of k-algebras $\{Ph^n(A), i_j^n\}_{n, 0 \le j \le n}$ has an inductive (direct) limit, $Ph^\infty A$, together with homomorphisms,

$$i_n : Ph^n(A) \longrightarrow Ph^\infty(A)$$

satisfying,

$$i_j^n \circ i_{n+1} = i_n, \; j = 0, 1, .., n.$$

Moreover, the family of derivations, $\{d_n\}_{0 \le n}$ define a unique derivation,

$$\delta : Ph^\infty(A) \longrightarrow Ph^\infty(A),$$

the Dirac derivation, such that,

$$i_n \circ \delta = d_n \circ i_{n+1},$$

and it is easy to see that this is a universal construction, i.e. any pair of a morphism,

$$i : A \longrightarrow B$$

and a derivation $\xi \in Der_k(B)$, factorizes via $Ph^\infty(A)$, and δ. Put

$$Ph^{(n)}(A) := im \; i_n \subseteq Ph^\infty(A)$$

Let for any associative algebra B, $Rep(B)$ denote the category of B-modules. The set of isomorphism-classes of B-modules is just a set, and the map induced by the obvious forgetful functor,

$$\omega : Rep(Ph^\infty(A)) \longrightarrow Rep(A),$$

is just a set-theoretical map, although having a well defined tangent map,

$$T_\omega : Ext^1_{Ph^\infty(A)}(V, V) \longrightarrow Ext^1_A(V, V).$$

As we shall see, assuming the algebra A of finite type, the set of simple finite dimensional A-modules form an algebraic scheme, $Simp_{\ne \infty}(A)$. Moreover,

Theorem 2.2.1 (Preparation). *The canonical morphism $i_0 : A \to Ph^\infty(A)$ parametrizes simple representations of A with fixed momentum, acceleration, and any number of higher order momenta.*

This should be understood in the following way. Consider, for any simple A-module V, the exact sequence,

$$0 \to End_A(V) \to End_k(V) \to Der_k(A, End_k(V)) \to Ext^1_A(V, V) \to 0.$$

Let $\rho : A \to End_k(V)$ define the structure of V, then any morphism $\rho_1 :$ $Ph(A) \to End_k(V)$ extending ρ, corresponds to a derivation, $\xi_1 : A \to$ $End_k(V)$, and therefore, via the maps in the exact sequence above, to a tangent vector, also called ξ_1, in the tangent space of the point $v \in$ $Simp(A)$ corresponding to V. So any such ξ_1 corresponds to a couple (v, ξ) of a point in $Simp(A)$, and an infinitesimal deformation of that point, i.e. a momentum. Any morphism $\rho_2 : Ph^2(A) \to End_k(V)$ extending ρ_1 corresponds therefore to the triple (v, ξ_1, ξ_2), corresponding to a point and a momentum, and to an infinitesimal deformation of this, etc. Since we have canonical morphisms $Ph^r(A) \to Ph^\infty(A)$, it is clear that any morphism $\xi : Ph^\infty(A) \to End_k(V)$ extending ρ, produces a sequence, of any order, of such tuples. A simple consequence of the definition of $Ph^\infty(A)$, that we identify all $i^n_q, q = 1, ..., n$, shows that the set of such morphisms ξ, extending a given structure-morphism, parametrizes the set of *formal curves* in $Simp(A)$ through v.

A fundamental problem of (our model of) physics, see the Introduction, can now be stated as follows: If we *prepare* an object so that we know its momentum, and its higher order momenta up to a certain order, what can we infer on its behavior in the future?

In our mathematical model, a *preparation* made on the A-module, the object, V, by fixing its structure as a $Ph^\infty(A)$-module, now forces it to change in the following way: The Dirac derivation $\delta \in Der_k(Ph^\infty(A))$ maps via the structure homomorphism of the module V,

$$\rho_V : Ph^\infty(A) \longrightarrow End_k(V)$$

to an element $\delta_V \in Der_k(Ph^\infty(A), End_k(V))$ and via the composition of the canonical linear maps,

$$Der_k(Ph^\infty(A), End_k(V)) \longrightarrow Ext^1_{Ph^\infty(A)}(V, V) \longrightarrow Ext^1_A(V, V)$$

to the element $\delta(V) \in Ext^1_A(V, V)$, i.e. to a tangent vector of $Simp_n(A)$ at the point, v, corresponding to V, see [18]. Suppose first that, $\delta(V)$ is 0. This means that δ_V in $Der_k(Ph^\infty(A), End_k(V))$ is an inner derivation given by an endomorphism $Q \in End_k(V)$, such that for every $f \in Ph^\infty(A)$, we find $\delta(f)(v) = (Qf - fQ)(v)$. This Q is the corresponding *Hamiltonian*, (or *Dirac operator* in the terminology of Connes, see [29]), and we have a situation that is very much like classical quantum mechanics, i.e. a set-up where the objects are represented by a fixed Hilbert space V and an algebra of *observables* $Ph^\infty(A)$ acting on it, with time, and therefore also energy, represented by a special Hamiltonian operator Q.

A system characterized by a $Ph^\infty(A)$-module V, for which $\delta(V) = 0$, will be called *stable* or *singular*. It is said to be in the *state* ψ if we have chosen an element $\psi \in V$. The Dirac derivation δ defines a Hamiltonian operator Q, (a Dirac operator), and time, i.e. δ, now push the state ψ into the state,

$$exp(\tau Q)(\psi) \in V,$$

corresponding to the isomorphism of the module V defined by the inner isomorphism of the algebra of observables, $Ph^\infty(A)$ defined by $U := exp(\tau\delta)$, whenever this is well defined. This is a well known situation i classical quantum mechanics, corresponding to the equivalence between the set-ups of Schrödinger and Heisenberg.

To treat the situation when $[\delta] \neq 0$, we first have to take a new look at non-commutative algebraic geometry, as developed in [17], [18], [19].

Chapter 3

Non-commutative Deformations and the Structure of the Moduli Space of Simple Representations

3.1 Non-commutative Deformations

In [16], [17] and [18], [19], we introduced non-commutative deformations of families of modules of non-commutative k-algebras, and the notion of *swarm* of right modules (or more generally of objects in a k-linear abelian category). Let for any associative k-algebra S, $\underline{a}_S$ be the category of S-valued associative k-algebras, the objects of which are the diagrams of k-algebras,

$$S \to^\iota R \to^\pi S$$

such that the composition of ι and π is the identity. In particular, $\underline{a}_r$ denotes the category of r-pointed k-algebras, i.e. $\underline{a}_s$, with $S = k^r$. Any such *r-pointed k-algebra* R is isomorphic to a k-algebra of $r \times r$-matrices $(R_{i,j})$.

The radical of R is the bilateral ideal $Rad(R) := ker\pi$, such that $R/Rad(R) \simeq k^r$. The dual k-vector space of $Rad(R)/Rad(R)^2$ is called the tangent space of R.

For $r = 1$, there is an obvious inclusion of categories

$$\underline{l} \subseteq \underline{a}_1$$

where $\underline{l}$, as usual, denotes the category of commutative local Artinian k-algebras with residue field k.

Fix a not necessarily commutative k-algebra A and consider a right A-module M. The ordinary deformation functor

$$Def_M : \underline{l} \to \underline{Sets}$$

is then defined. Assuming $Ext^i_A(M, M)$ has finite k-dimension for $i = 1, 2$, it is well known, see [28], or [16], that Def_M has a pro-representing hull H, *the formal moduli of M.* Moreover, the tangent space of H is isomorphic to $Ext_A^1(M, M)$, and H can be computed in terms of $Ext_A^i(M, M)$, $i = 1, 2$ and their *matric Massey products*, see [15], [16], [21].

In the general case, consider a finite family $\mathbf{V} = \{V_i\}_{i=1}^r$ of right A-modules. Assume that,

$$dim_k Ext_A^1(V_i, V_j) < \infty.$$

Any such family of A-modules will be called a *swarm*. We shall define a *deformation functor*,

$$Def_{\mathbf{V}} : \underline{a}_r \to \underline{Sets}$$

generalizing the functor Def_M above. Given an object $\pi : R = (R_{i,j}) \to k^r$ of $\underline{a}_r$, consider the k-vector space and left R-module $(R_{i,j} \otimes_k V_j)$. It is easy to see that $End_R((R_{i,j} \otimes_k V_j)) \simeq (R_{i,j} \otimes_k Hom_k(V_i, V_j))$. Clearly π defines a k-linear and left R-linear map,

$$\pi(R) : (R_{i,j} \otimes_k V_j) \to \oplus_{i=1}^r V_i,$$

inducing a homomorphism of R-endomorphism rings,

$$\tilde{\pi}(R) : (R_{i,j} \otimes_k Hom_k(V_i, V_j)) \to \oplus_{i=1}^r End_k(V_i).$$

The right A-module structure on the $V_i's$ is defined by a homomorphism of k-algebras, $\eta_0 : A \to \oplus_{i=1}^r End_k(V_i)$. Let

$$Def_{\mathbf{V}}(R) \in \underline{Sets}$$

be the set of isoclasses of homomorphisms of k-algebras,

$$\eta' : A \to (R_{i,j} \otimes_k Hom_k(V_i, V_j))$$

such that,

$$\tilde{\pi}(R) \circ \eta' = \eta_0,$$

where the equivalence relation is defined by inner automorphisms in the k-algebra $(R_{i,j} \otimes_k Hom_k(V_i, V_j))$ inducing the identity on $\oplus_{i=1}^r End_k(V_i)$. One easily proves that $Def_{\mathbf{V}}$ has the same properties as the ordinary deformation functor and we prove the following, see [15], [16], [21].

Theorem 3.1.1. *The functor $Def_{\mathbf{V}}$ has a pro-representable hull, i.e. an object H of the category of pro-objects $\underline{\hat{a}}_r$ of $\underline{a}_r$, together with a versal family,*

$$\tilde{\mathbf{V}} = \{(H_{i,j} \otimes V_j)\}_{i=1}^r \in \varprojlim_{n \geq 1} Def_{\mathbf{V}}(H/\mathfrak{m}^n),$$

where $\mathfrak{m} = Rad(H)$, *such that the corresponding morphism of functors on* $\underline{a}_r$,

$$\kappa : Mor(H, -) \to Def_{\mathbf{V}}$$

defined for $\phi \in Mor(H, R)$ *by* $\kappa(\phi) = R \otimes_\phi \tilde{V}$, *is smooth, and an isomorphism on the tangent level. Moreover, H is uniquely determined by a set of matric Massey products defined on subspaces,*

$$D(n) \subseteq Ext^1(V_i, V_{j_1}) \otimes \cdots \otimes Ext^1(V_{j_{n-1}}, V_k),$$

with values in $Ext^2(V_i, V_k)$. *The right action of A on $\tilde{V}$ defines a homomorphism of k-algebras, the versal family,*

$$\eta : A \longrightarrow O(\mathbf{V}) := End_H(\tilde{V}) = (H_{i,j} \otimes Hom_k(V_i, V_j)),$$

and the k-algebra $O(\mathbf{V})$ acts on the family of A-modules $\mathbf{V} = \{V_i\}$, extending the action of A. If $dim_k V_i < \infty$, for all $i = 1, ..., r$, the operation of associating $(O(\mathbf{V}), \mathbf{V})$ to $(A, \mathbf{V})$ is a closure operation.

3.2 The O-construction

Moreover, in [16] we prove the crucial result,

Theorem 3.2.1 (A generalized Burnside theorem). *Let A be a finite dimensional k-algebra, k an algebraically closed field. Consider the family* $\mathbf{V} = \{V_i\}_{i=1}^r$ *of all simple A-modules, then*

$$\eta : A \longrightarrow O(\mathbf{V}) = (H_{i,j} \otimes Hom_k(V_i, V_j))$$

is an isomorphism.

This result may be substantially generalized. In fact, let S be any finite dimensional k-algebra, and consider the category $\underline{a}_S$. An object is a diagram,

$$R := (S \to^\iota R \to^\pi S)$$

where R is a finite dimensional k-algebra, and the composition $\iota\pi = id$. Morphisms are the obvious commutative diagrams. Suppose we are given a homomorphism of k-algebras,

$$\rho_0 : A \to End_S(V),$$

where V is finite dimensional left S-module. For any object, $R \in \underline{a}_S$, we may consider the left R-module $R \otimes_S V$, and the homomorphism of k-algebras,

$$\pi_* : End_R(R \otimes_S V) \to End_S(V).$$

The same arguments as above, then proves the more general result,

Theorem 3.2.2 (The O-construction Theorem). *Let V be any left S- and right A-module, and consider the functor,*

$$Def_{(sV_A)} : \underline{a}_S \to \underline{Sets},$$

defined by,

$$Def_{(sV_A)}(R) = \{\rho \in Mor_k(A, End_R(R \otimes_S V)) | \phi\pi_* = \rho_0\}/\omega,$$

where the equivalence relation ω is defined by inner automorphisms in the k-algebra $End_R(R \otimes_S V)$, inducing the identity on $End_S(V)$. This functor has a pro-representable hull, i.e. an object H, of the category of pro-objects $\underline{\hat{a}}_S$ of $\underline{a}_S$, with projection, $\hat{\pi} : H \to S$, together with a versal family,

$$\tilde{V} \in \varprojlim_{n \geq 1} Def_{sV_A}(H/\mathfrak{m}^n),$$

where $\mathfrak{m} = Rad(H) := ker\hat{\pi}$, such that the corresponding morphism of functors on $\underline{a}_r$,

$$\kappa : Mor(H, -) \to Def_\mathbf{V}$$

defined for $\phi \in Mor(H, R)$ by $\kappa(\phi) = R \otimes_\phi \tilde{V}$, is smooth, and an isomorphism on the tangent level. Moreover, H is uniquely determined by a set of matric Massey products with values in $Ext^2(V, V)$. The right action of A on $\tilde{V}$ defines a homomorphism of k-algebras, the versal family,

$$\eta : A \longrightarrow O(sV_A) := End_H(\tilde{V}).$$

The k-algebra $O := O(sV_A)$ acts on the module V, extending the right-action of A, commuting with the S-action. If $dim_k V < \infty$, the operation of associating $(O(sV_A), sV_O)$ to (A, sV_A), is a closure operation.

Notice that the proof of the closure property of the O-construction, as proposed in [18], is incomplete, and should be replaced with the above. Details will occur in a projected book.

We also proved that there exists, in the non-commutative deformation theory, an obvious analogy to the notion of pro-representing (modular) substratum H_0 of the formal moduli H, see [14] and [22]. The tangent space of H_0 is determined by a family of subspaces

$$Ext_0^1(V_i, V_j) \subseteq Ext_A^1(V_i, V_j), \qquad i \neq j$$

the elements of which should be called the *almost split extensions* (sequences) relative to the family $\mathbf{V}$, and by a subspace,

$$T_0(\Delta) \subseteq \prod_i Ext_A^1(V_i, V_i)$$

which is the tangent space of the deformation functor of the full subcategory of the category of A-modules generated by the family $\mathbf{V} = \{V_i\}_{i=1}^r$, see [15]. If $\mathbf{V} = \{V_i\}_{i=1}^r$ is the set of all indecomposables of some Artinian k-algebra A, we show that the above notion of *almost split sequence* coincides with that of Auslander, see [26].

3.3 Iterated Extensions

In [16], we consider the general problem of classification of *iterated extensions* of a family of modules $\mathbf{V} = \{V_i\}_{i=1}^r$, and the corresponding classification of *filtered modules* with graded components in the family $\mathbf{V}$, and *extension type* given by a *directed representation graph* Γ, see Chapter 4.

The main result is the following result, see [18],

Proposition 3.3.1. *Let A be any k-algebra, $\mathbf{V} = \{V_i\}_{i=1}^r$ any swarm of A-modules, i.e. such that,*

$$\dim_k Ext_A^1(V_i, V_j) < \infty \quad \text{for all } i, j = 1, \ldots, r.$$

(i): Consider an iterated extension E of $\mathbf{V}$, with representation graph Γ. Then there exists a morphism of k-algebras

$$\phi : H(\mathbf{V}) \to k[\Gamma]$$

such that

$$E \simeq k[\Gamma] \otimes_\phi \tilde{V}$$

as right A-modules.

(ii): The set of equivalence classes of iterated extensions of $\mathbf{V}$ with representation graph Γ, is a quotient of the set of closed points of the affine algebraic variety

$$\underline{A}[\Gamma] = Mor(H(\mathbf{V}), k[\Gamma])$$

(iii): There is a versal family $\tilde{V}[\Gamma]$ of A-modules defined on $\underline{A}[\Gamma]$, containing as fibers all the isomorphism classes of iterated extensions of $\mathbf{V}$ with representation graph Γ.

Let $Mod_A^{\mathbf{V}}$ denote the full subcategory of Mod_A generated by the iterated extensions of of $\mathbf{V}$. As usual denote by $H := H(\mathbf{V}$ the formal moduli of $\mathbf{V}$, Then we have the following structure theorem, generalizing a result, of Beilinson,

Theorem 3.3.2. *Let A be any k-algebra, and fix a swarm, $\mathbf{V} = \{V_i\}_{i=1}^r$, of A-modules, then there exists a functor,*

$$\kappa : Mod_{H(\mathbf{V})} \to Mod_A^{\mathbf{V}}$$

which is an isomorphism on equivalence-classes of objects, and monomorphic on morphisms. If $\mathbf{V}$ consists of simple modules, then κ is an equivalence.

Proof.   Any right $H(\mathbf{V})$-module M, is a k^r-module, so it can be decomposed as, $M = \oplus M_i$, where $M_i := Me_i$. The structure map is therefore given as,

$$\rho_0 : H(\mathbf{V}) \rightarrow End_k(M) = (Hom_k(M_i, M_j)).$$

Here, ρ_0 maps $H_{i,j}$ into $Hom_k(M_i, M_j)$, and therefore the formal family may be decomposed to give us the following k-algebra homomorphisms,

$$\rho : A \rightarrow (H_{i,j} \otimes Hom_k(V_i, V_j)) \rightarrow (Hom_k(M_i, M_j) \otimes_k Hom_k(V_i, V_j))$$
$$= (Hom_k(M_i \otimes_k V_i, M_j \otimes V_j)) = End_k(W).$$

Here $W = \oplus_{i=1}^r (M_i \otimes V_i)$, and $\kappa(M) := W$. Since $k[\Gamma] \subset End_k(P)$, where P is $k[\Gamma]$, as $k[\Gamma]$-module, the first part of the theorem follows from Proposition (3.3.1). The rest is more or less evident. $\square$

Notice that in the literature one finds the notions *cluster* and *mutations*, both closely related to iterated extensions of modules over *quivers*, or as we termed it, presheaves defined on partially ordered sets, and categories, see e.g. [13].

3.4   Non-commutative Schemes

To any, not necessarily finite, swarm $\underline{c} \subset \underline{mod}(A)$ of right-A-modules, we may use the above O-construction, to associated to $\underline{c}$ a k-algebra, see [17] and [18],

$$O(\underline{c}, \pi) = \varprojlim_{\underline{c}_0 \subset |\underline{c}|} O(_S V_A),$$

where $S = k[\underline{c}_0]$, is the k-algebra of the quiver associated to $\underline{c}_0$, where $\{V_i\}_{i=1}^r = |\underline{c}_0|$, and where $V := \oplus_i^r V_i$. There is a natural k-algebra homomorphism,

$$\eta(\underline{c}) : A \longrightarrow O(\underline{c})$$

with the property that the A-module structure on $\underline{c}$ is extended to an **O**-module structure in an *optimal* way. We then defined an *affine noncommutative scheme* of right A-modules to be a swarm $\underline{c}$ of right A-modules, such that $\eta(\underline{c})$ is an isomorphism. In particular we considered, for finitely generated k-algebras, the swarm $Simp_{<\infty}^*(A)$ consisting of the finite dimensional simple A-modules, and the *generic* point A, together with all morphisms between them. The fact that this is a swarm, i.e. that for all objects $V_i, V_j \in Simp_{<\infty}$ we have $dim_k Ext_A^1(V_i, V_j) < \infty$, is easily proved.

For *geometric k*-algebras, see [18], we have proved the following result, see (4.1), loc.cit.,

Theorem 3.4.1. *Let A be a geometric k-algebra, then the natural homomorphism,*

$$\eta(Simp^*(A)) : A \longrightarrow \mathbf{O}_\pi(Simp^*_{<\infty}(A))$$

is an isomorphism, i.e. $Simp^*_{<\infty}(A)$ is a scheme for A.

In particular, the moduli space, $Simp^*_{<\infty}(k < x_1, x_2, ..., x_d >)$, is a scheme for $k < x_1, x_2, ..., x_d >$.

To analyze the local structure of $Simp_n(A)$, we need the following, see [18], (3.23),

Lemma 3.4.2. *Let $\mathbf{V} = \{V_i\}_{i=1,..,r}$ be a finite subset of $Simp_{<\infty}(A)$, then the morphism of k-algebras,*

$$A \to O(\mathbf{V}) = (H_{i,j} \otimes_k Hom_k(V_i, V_j))$$

is topologically surjective.

Proof. Since the simple modules V_i, $i = 1, .., r$ are distinct, there is an obvious surjection, $\eta_0 : A \to \prod_{i=1,..,r} End_k(V_i)$. Put $\mathfrak{r} = ker\eta_0$, and consider for $m \geq 2$ the finite-dimensional k-algebra, $B := A/\mathfrak{r}^m$. Clearly $Simp(B) = \mathbf{V}$, so that by the generalized Burnside theorem, see [16], (2.6), we find, $B \simeq O^B(\mathbf{V}) := (H^B_{i,j} \otimes_k Hom_k(V_i, V_j))$. Consider the commutative diagram,

$$
\begin{array}{ccc}
A & \longrightarrow & (H^A_{i,j} \otimes_k Hom_k(V_i, V_j)) =: O^A(\mathbf{V}) \\
\downarrow & & \downarrow \qquad\qquad\qquad \searrow \\
B & \longrightarrow & (H^B_{i,j} \otimes_k Hom_k(V_i, V_j)) \xrightarrow{\ \alpha\ } O^A(\mathbf{V})/\mathfrak{m}^m
\end{array}
$$

where all morphisms are natural. In particular α exists since $B = A/\mathfrak{r}^m$ maps into $O^A(\mathbf{V})/rad^m$, and therefore induces the morphism α commuting with the rest of the morphisms. Consequently α has to be surjective, and we have proved the contention. $\qquad\square$

3.4.1 Localization, Topology and the Scheme Structure on Simp(A)

Let $s \in A$, and consider the subset $D(s) = \{V \in Simp(A) | \rho(s)^{-1} \in End_k(V)\}$ of $Simp(A)$.

The Jacobson topology on $Simp(A)$ is the topology with basis $\{D(s)|\ s \in A\}$. It is clear that the natural morphism,

$$\eta : A \to O(D(s), \pi)$$

maps s into an invertible element of $O(D(s), \pi)$. Therefore we may define the localization $A_{\{s\}}$ of A, as the k-algebra generated in $O(D(s), \pi)$ by $im\ \eta$ and the inverse of $\eta(s)$. This furnishes a general method of localization with all the properties one would wish. And in this way we also find a canonical (pre)sheaf, **O** defined on $Simp(A)$.

Definition 3.4.3. When the k-algebra A is geometric, such that $Simp^*(A)$ is a scheme for A, we shall refer to the presheaf **O**, defined above on the Jacobson topology, as the structure presheaf of the scheme $Simp(A)$.

We shall now see that the Jacobson topology on $Simp(A)$, restricted to each $Simp_n(A)$ is the Zariski topology for a classical scheme-structure.

Recall first that a *standard n-commutator relation* in a k-algebra A is a relation of the type,

$$[a_1, a_2, ..., a_{2n}] := \sum_{\sigma \in \Sigma_{2n}} sign(\sigma) a_{\sigma(1)} a_{\sigma(2)} ... a_{\sigma(2n)} = 0$$

where $\{a_1, a_2, ..., a_{2n}\}$ is a subset of A. Let $I(n)$ be the two-sided ideal of A generated by the subset,

$$\{[a_1, a_2, ..., a_{2n}]|\ \{a_1, a_2, ..., a_{2n}\} \subset A\}.$$

Consider the canonical homomorphism,

$$p_n : A \longrightarrow A/I(n) =: A(n).$$

It is well known that any homomorphism of k-algebras,

$$\rho : A \longrightarrow End_k(k^n) =: M_n(k)$$

factors through p_n, see e.g. [4].

Corollary 3.4.4. *(i). Let $V_i, V_j \in Simp_{\leq n}(A)$ and put $\mathbf{r} = \mathbf{m}_{V_i} \cap \mathbf{m}_{V_j}$. Then we have, for $m \geq 2$,*

$$Ext^1_A(V_i, V_j) \simeq Ext^1_{A/\mathbf{r}^m}(V_i, V_j)$$

(ii). Let $V \in Simp_n(A)$. Then,

$$Ext^1_A(V, V) \simeq Ext^1_{A(n)}(V, V)$$

Proof. (i) follows directly from Lemma (3.4.2). To see (ii), notice that $Ext_A^1(V, V) \simeq Der_k(A, End_k(V))/Triv \simeq Der_k(A(n), End_k(V))/Triv \simeq Ext_{A(n)}^1(V, V)$. The second isomorphism follows from the fact that any derivation maps a standard n-commutator relation into a sum of standard n-commutator relations. $\square$

Example 3.1. Notice that, for distinct $V_i, V_j \in Simp_{\leq n}(A)$, we may well have,

$$Ext_A^1(V_i, V_j) \neq Ext_{A(n)}^1(V_i, V_j).$$

In fact, consider the matrix k-algebra,

$$A = \begin{pmatrix} k[x] & k[x] \\ 0 & k[x] \end{pmatrix},$$

and let $n = 1$. Then $A(1) = k[x] \oplus k[x]$. Put $V_1 = k[x]/(x) \oplus (0), V_2 = (0) \oplus k[x]/(x)$, then it is easy to see that,

$$Ext_A^1(V_i, V_j) = k, \ Ext_{A(1)}^1(V_i, V_j) = 0, i \neq j,$$

but,

$$Ext_A^1(V_i, V_i) = Ext_{A(1)}^1(V_i, V_i) = k, i = 1, 2.$$

Lemma 3.4.5. *Let B be a k-algebra, and let V be a vector space of dimension n, such that the k-algebra $B \otimes End_k(V)$ satisfies the standard n-commutator-relations, i.e. such that the ideal, $I(n) \subset B \otimes End_k(V)$ generated by the standard n-commutators $[x_1, x_2, .., x_{2n}]$, $x_i \in B \otimes End_k(V)$, is zero. Then B is commutative.*

Proof. In fact, if $b_1, b_2 \in B$ is such that $[b_1, b_2] \neq 0$, then the obvious n-commutator,

$$(b_1 e_{1,1})(b_2 e_{1,1}) e_{1,2} e_{2,2} \dots e_{n-1,n} \cdot e_{n,n} - (b_2 e_{1,1})(b_1 e_{1,1}) e_{1,2} e_{2,2} \dots e_{n-1,n} \cdot e_{n,n}$$

is different from 0. Here $e_{i,j}$ is the $n \times n$ matrix with all elements equal to 0, except the one in the (i, j) position, where the element is equal to 1. $\square$

Lemma 3.4.6. *If A is a finite type k-algebra, then any $V \in Simp_n(A)$ is an $A(n)$-module. Let $\mathbf{V} \subset Simp_n(A)$ be a finite family, then $H^{A(n)}(\mathbf{V})$ is commutative. In particular,*

- *$Ext_{A(n)}(V_i, V_j) = 0$, for $V_i \neq V_j$*
- *$H^{A(n)}(V) \simeq H^A(V)^{com} := H(V)/[H(V), H(V)]$.*

Proof.   Since

$$A(n) \to O(\mathbf{V}) \simeq M_n(H^{A(n)}(\mathbf{V}))$$

is topologically surjective, we find using (Lemma 3.4.9), that $H^{A(n)}(\mathbf{V})$ is commutative. This implies (1) and the commutativity of $H^{A(n)}(V)$. Consider for $V \in Simp_n(A)$, the natural commutative diagram of homomorphisms of k-algebras,

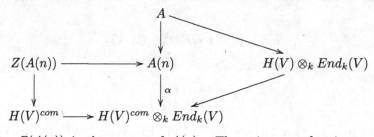

where $Z(A(n))$ is the center of $A(n)$. The existence of α is a consequence of the ideal $I(n)$ of A mapping to zero in $H(V)^{com} \otimes_k End_k(V) \simeq M_n(H(V)^{com})$. Therefore there are natural morphisms of formal moduli,

$$H^A(V) \to H^{A(n)}(V) \to H^A(V)^{com} \to H^{A(n)}(V)^{com}.$$

Since $H^{A(n)}(V) = H^{A(n)}(V)^{com}$ the composition,

$$H^{A(n)}(V) \to H^A(V)^{com} \to H^{A(n)}(V)^{com},$$

must be an isomorphism. Since, by Corollary (3.4.4), the tangent spaces of $H^{A(n)}(V)$ and $H^A(V)$ are isomorphic, the lemma is proved.   □

Corollary 3.4.7. *Let* $A = k < x_1, .., x_d >$ *be the free* k-*algebra on* d *symbols, and let* $V \in Simp_n(A)$. *Then*

$$H^A(V)^{com} \simeq H^{A(n)}(V) \simeq k[[t_1, ..., t_{(d-1)n^2+1}]]$$

This should be compared with the results of [24], see also [4].
   In general, the natural morphism,

$$\eta(n) : A(n) \to \prod_{V \in Simp_n(A)} H^{A(n)}(V) \otimes_k End_k(V)$$

is not an injection, as it follows from the following,

Example 3.2. Let

$$A = \begin{pmatrix} k & k & k \\ k & k & k \\ 0 & 0 & k \end{pmatrix}.$$

The ideal $I(2)$ is generated by $[e_{1,1}, e_{1,2}, e_{2,2}, e_{2,3}] = e_{1,3}$. So

$$A(2) = \begin{pmatrix} k & k & k \\ k & k & k \\ 0 & 0 & k \end{pmatrix} / \begin{pmatrix} 0 & 0 & k \\ 0 & 0 & k \\ 0 & 0 & 0 \end{pmatrix} \simeq M_2(k) \oplus M_1(k).$$

However,

$$\prod_{V \in Simp_2(A)} H^{A(2)}(V) \otimes_k End_k(V) \simeq M_2(k),$$

therefore $ker\, \eta(2) = M_1(k) = k$.

Let $O(n)$, be the image of $\eta(n)$, then,

$$O(n) \subseteq \prod_{V \in Simp_n(A)} H^{A(n)}(V) \otimes_k End_k(V)$$

and for every $V \in Simp_n(A)$,

$$H^{O(n)}(V) \simeq H^{A(n)}(V).$$

Put $B = \prod_{V \in Simp_n(A)} H^{A(n)}(V)$. Choosing bases in all $V \in Simp_n(A)$, then

$$\prod_{V \in Simp_n(A)} H^{A(n)}(V) \otimes_k End_k(V) \simeq M_n(B),$$

Let $x_i \in A, i = 1, ..., d$ be generators of A, and consider their images $(x_{p,q}^i) \in M_n(B)$. Now, B is commutative, so the k-sub-algebra $C(n) \subset B$ generated by the elements $\{x_{p,q}^i\}_{i=1,..,d;\ p,q=1,..,n}$ is commutative. We have an injection,

$$O(n) \rightarrow M_n(C(n)),$$

and for all $V \in Simp_n(A)$, with a chosen basis, there is a natural composition of homomorphisms of k-algebras,

$$\alpha : M_n(C(n)) \rightarrow M_n(H^{A(n)}(V)) \rightarrow End_k(V),$$

inducing a corresponding composition of homomorphisms of the centers,

$$Z(\alpha) : C(n) \rightarrow H^{A(n)}(V) \rightarrow k$$

This sets up a set theoretical injective map,

$$t : Simp_n(A) \longrightarrow Max(C(n)),$$

defined by $t(V) := ker Z(\alpha)$.

Since $A(n) \to H^{A(n)}(V) \otimes_k End_k(V)$ is topologically surjective, $H^{A(n)}(V) \otimes_k End_k(V)$ is topologically generated by the images of x_i, $i = 1, ..., d$. It follows that we have a surjective homomorphism,

$$\hat{C}(n)_{t(V)} \to H^{A(n)}(V).$$

Categorical properties implies, that there is another natural morphism,

$$H^{A(n)}(V) \to \hat{C}(n)_{t(V)},$$

which composed with the former is an automorphism of $H^{A(n)}(V)$. Since

$$M_n(C(n)) \subseteq \prod_{V \in Simp_n(A)} H^{A(n)}(V) \otimes_k End_k(V),$$

it follows that for $\mathfrak{m}_v \in Max(C(n))$, corresponding to $V \in Simp_n(A)$, the finite dimensional k-algebra $M_n(C(n)/\mathfrak{m}_v{}^2)$ sits in a finite dimensional quotient of some,

$$\prod_{V \in \mathbf{V}} H^{A(n)}(V) \otimes_k End_k(V).$$

where $\mathbf{V} \subset Simp_n(A)$ is finite. However, by Lemma (2.5), the morphism,

$$A(n) \longrightarrow \prod_{V \in \mathbf{V}} H^{A(n)}(V) \otimes_k End_k(V)$$

is topologically surjectiv. Therefore the morphism,

$$A(n) \longrightarrow M_n(C(n)/\mathfrak{m}_v{}^2)$$

is surjectiv, implying that the map

$$H^{A(n)}(V) \to \hat{C}(n)_{\mathfrak{m}_v},$$

is surjectiv, and consequently, $H^{A(n)}(V) \simeq \hat{C}(n)_{\mathfrak{m}_v}$.

We now have the following theorem, see Chapter VIII, §2, of the book [25], where part of this theorem is proved.

Theorem 3.4.8. *Let* $V \in Simp_n(A)$, *correspond to the point* $\mathfrak{m}_v \in Simp_1(C(n))$.

(i) There exist a Zariski neighborhood U_v *of* v *in* $Simp_1(C(n))$ *such that any closed point* $\mathfrak{m}'_v \in U$ *corresponds to a unique point* $V' \in Simp_n(A)$.

Let $U(n)$ *be the open subset of* $Simp_1(C(n))$, *the union of all* U_v *for* $V \in Simp_n(A)$.

(ii) $O(n)$ *defines a non-commutative structure sheaf* $\mathbf{O}(n) := \mathbf{O}_{U(n)}$ *of Azumaya algebras on the topological space* $U(n)$ *(Jacobson topology).*

(iii) The center $\mathbf{S}(n)$ *of* $\mathbf{O}(n)$, *defines a scheme structure on* $Simp_n(A)$.

(iv) The versal family of n-dimensional simple modules, $\tilde{V} := C(n) \otimes_k V$, over $Simp_n(A)$, is defined by the morphism,

$$\tilde{\rho} : A \to O(n) \subseteq End_{C(n)}(C(n)) \otimes_k V) \simeq M_n(C(n)).$$

(v) The trace ring $Tr\tilde{\rho} \subseteq \mathbf{S}(n) \subseteq C(n)$ defines a commutative affine scheme structure on $Simp_n(A)$. Moreover, there is a morphism of schemes,

$$\kappa : U(n) \longrightarrow Simp_n(A),$$

such that for any $v \in U(n)$,

$$H^{A(n)}(V) \simeq \hat{\mathbf{S}}(n)_{\kappa(v)} \simeq (\hat{T}r\tilde{\rho})_{\kappa(v)} \simeq \hat{C}(n)_v$$

Proof. Let $\rho : A \longrightarrow End_k(V)$ be the surjective homomorphism of k-algebras, defining $V \in Simp_n(A)$. Let, as above $e_{i,j} \in End_k(V)$ be the elementary matrices, and pick $y_{i,j} \in A$ such that $\rho(y_{i,j}) = e_{i,j}$. Let us denote by σ the cyclical permutation of the integers $\{1, 2, ..., n\}$, and put,

$$s_k := [y_{\sigma^k(1),\sigma^k(2)}, y_{\sigma^k(2),\sigma^k(2)}, y_{\sigma^k(2),\sigma^k(3)} \cdots y_{\sigma^k(n),\sigma^k(n)}],$$

$$s := \sum_{k=0,1,..,n-1} s_k \in A.$$

Clearly, $s \in I(n-1)$.

Since $[e_{\sigma^k(1),\sigma^k(2)}, e_{\sigma^k(2),\sigma^k(2)}, e_{\sigma^k(2),\sigma^k(3)} \cdots e_{\sigma^k(n),\sigma^k(n)}] = e_{\sigma^k(1),\sigma^k(n)} \in End_k(V)$, $\rho(s) := \sum_{k=0,1,..,n-1} \rho(s_k) \in End_k(V)$ is the matrix with non-zero elements, equal to 1, only in the $(\sigma^k(1), \sigma^k(n))$ position, so the determinant of $\rho(s)$ must be $+1$ or -1. The determinant $det(s) \in C(n)$ is therefore nonzero at the point $v \in Spec(C(n))$ corresponding to V. Put $U = D(det(s)) \subset Spec(C(n))$, and consider the localization $O(n)_{\{s\}} \subseteq M_n(C(n)_{\{det(s)\}})$, the inclusion following from general properties of the localization. Now, any closed point $v' \in U$ corresponds to a n-dimensional representation of A, for which the element $s \in I(n-1)$ is invertible. But then this representation cannot have a $m < n$ dimensional quotient, so it must be simple.

Since $s \in I(n-1)$, the localized k-algebra $O(n)_{\{s\}}$ does not have any simple modules of dimension less than n, and no simple modules of dimension $> n$. In fact, for any finite dimensional $O(n)_{\{s\}}$-module V, of dimension m, the image $\hat{s}$ of s in $End_k(V)$ must be invertible. However, the inverse $\hat{s}^{-1}$ must be the image of a polynomial (of degree $m-1$) in s. Therefore, if V is simple over $O(n)_{\{s\}}$, i.e. if the homomorphism $O(n)_{\{s\}} \to End_k(V)$ is surjective, V must also be simple over A. Since now $s \in I(n-1)$, it follows that

$m \geq n$. If $m > n$, we may construct, in the same way as above an element in $I(n)$ mapping into a nonzero element of $End_k(V)$. Since, by construction, $I(n) = 0$ in $A(n)$, and therefore also in $O(n)_{\{s\}}$, we have proved what we wanted. By a theorem of M. Artin, see [1], $O(n)_{\{s\}}$ must be an Azumaya algebra with center, $S(n)_{\{s\}} := Z(O(n)_{\{s\}})$. Therefore $O(n)$ defines a presheaf $\mathbf{O}(n)$ on $U(n)$, of Azumaya algebras with center $\mathbf{S}(n) := Z(\mathbf{O}(n))$. Clearly, any $V \in Simp_n(A)$, corresponding to $\mathfrak{m}_v \in Max(C(n))$ maps to a point $\kappa(v) \in Simp(\mathbf{O}(n))$. Let $\mathfrak{m}_{\kappa(v)}$ be the corresponding maximal ideal of $\mathbf{S}(n)$. Since $O(n)$ is locally Azumaya, it follows that,

$$\hat{\mathbf{S}}(n)_{\mathfrak{m}_{\kappa(v)}} \simeq H^{O(n)}(V) \simeq H^{A(n)}(V).$$

The rest is clear.   □

$Spec(C(n))$ is, in a sense, a compactification of $Simp_n(A)$. It is, however, not the correct *completion* of $Simp_n(A)$. In fact, the points of $Spec(C(n)) - Simp_n(A)$ may correspond to semi-simple modules, which do not deform into simple n-dimensional modules. We shall return to the study of the (notion of) completion, together with the degeneration processes that occur, at *infinity* in $Simp_n(A)$.

Example 3.3. Let us check the case of $A = k < x_1, x_2 >$, the free non-commutative k-algebra on two symbols. First, we shall compute $Ext_A^1(V, V)$ for a particular $V \in Simp_2(A)$, and find a basis $\{t_i^*\}_{i=1}^5$, represented by derivations $\partial_i := \partial_i(V) \in Der_k(A, End_k(V))$, i=1,2,3,4,5. This is easy, since for any two A-modules V_1, V_2, we have the exact sequence,

$$0 \to Hom_A(V_1, V_2) \to Hom_k(V_1, V_2) \to Der_k(A, Hom_k(V_1, V_2))$$
$$\to Ext_A^1(V_1, V_2) \to 0$$

proving that, $Ext_A^1(V_1, V_2) = Der_k(A, Hom_k(V_1, V_2))/Triv$, where $Triv$ is the sub-vector space of trivial derivations. Pick $V \in Simp_2(A)$ defined by the homomorphism $A \to M_2(k)$ mapping the generators x_1, x_2 to the matrices

$$X_1 := \begin{pmatrix} 0 & 1 \\ 0 & 0 \end{pmatrix} =: e_{1,2}, \quad X_2 := \begin{pmatrix} 0 & 0 \\ 1 & 0 \end{pmatrix} =: e_{2,1}.$$

Notice that

$$X_1 X_2 = \begin{pmatrix} 1 & 0 \\ 0 & 0 \end{pmatrix} =: e_{1,1} = e_1, \quad X_2 X_1 = \begin{pmatrix} 0 & 0 \\ 0 & 1 \end{pmatrix} =: e_{2,2} = e_2,$$

and recall also that for any 2×2-matrix $(a_{p,q}) \in M_2(k)$, $e_i(a_{p,q})e_j = a_{i,j}e_{i,j}$. The trivial derivations are generated by the derivations $\{\delta_{p,q}\}_{p,q=1.2}$, defined by,

$$\delta_{p,q}(x_i) = X_i e_{p,q} - e_{p,q} X_i.$$

Clearly $\delta_{1,1} + \delta_{2,2} = 0$. Now, compute and show that the derivations ∂_i, $i = 1, 2, 3, 4, 5$, defined by,

$$\partial_i(x_1) = 0, \text{for } i = 1, 2, \ \partial_i(x_2) = 0, \text{for } i = 4, 5,$$

by,

$$\partial_1(x_2) = e_{1,1}, \partial_2(x_2) = e_{1,2}, \partial_3(x_1) = e_{1,2}, \partial_4(x_1) = e_{2,2}, \partial_5(x_1) = e_{2,1}$$

and by,

$$\partial_3(x_2) = e_{2,1},$$

form a basis for $Ext_A^1(V, V) = Der_k(A, End_k(V))/Triv$. Since $Ext_A^2(V, V) = 0$ we find $H(V) = k << t_1, t_2, t_3, t_4, t_5 >>$ and so $H(V)^{com} \simeq k[[t_1, t_2, t_3, t_4, t_5]]$. The formal versal family $\hat{V}$, is defined by the actions of x_1, x_2, given by,

$$X_1 := \begin{pmatrix} 0 & 1+t_3 \\ t_5 & t_4 \end{pmatrix}, \ X_2 := \begin{pmatrix} t_1 & t_2 \\ 1+t_3 & 0 \end{pmatrix}.$$

One checks that there are polynomials of X_1, X_2 which are equal to $t_i e_{p,q}$, modulo the ideal $(t_1, .., t_5)^2 \subset H(V)$, for all $i, p, q = 1, 2$. This proves that $\hat{C}(2)_v$ must be isomorphic to $H(V)$, and that the composition,

$$A \longrightarrow A(2) \longrightarrow M_2(C(2)) \subset M_2(H(V)))$$

is topologically surjective. By the construction of $C(n)$ this also proves that

$$C(2) \simeq k[t_1, t_2, t_3, t_4, t_5].$$

locally in a Zariski neighborhood of the origin. Moreover, the Formanek center, see [4], in this case is cut out by the single equation:

$$f := det[X_1, X_2] = -((1+t_3)^2 - t_2 t_5)^2 + (t_1(1+t_3) + t_2 t_4)(t_4(1+t_3) + t_1 t_5).$$

Computing, we also find the following formulas,

$$trX_1 = t_4, \ trX_2 = t_1,$$
$$detX_1 = -t_5 - t_3 t_5, \ detX_2 = -t_2 - t_2 t_3,$$
$$tr(X_1 X_2) = (1+t_3)^2 + t_2 t_5$$

so the *trace ring* of this family is

$$k[t_1, t_2 + t_2 t_3, 1 + 2t_3 + t_3^2 + t_2 t_5, t_4, t_5 + t_3 t_5] =: k[u_1, u_2, ..., u_5],$$

with,

$$u_1 = t_1, \ u_2 = (1+t_3)t_2, \ u_3 = (1+t_3)^2 + t_2 t_5, \ u_4 = t_4, \ u_5 = (1+t_3)t_5,$$

and $f = -u_3^2 + 4u_2 u_5 + u_1 u_3 u_4 + u_1^2 u_5 + u_2 u_4^2$. Moreover, the $k[\underline{t}]$ is algebraic over $k[\underline{u}]$, with discriminant, $\Delta := 4u_2 u_5(u_3^2 - 4u_2 u_5) = 4(1+t_3)^2 t_2 t_5((1+t_3)^2 - t_2 t_5)^2$. From this follows that there is an étale covering

$$\mathbf{A}^5 - V(f\Delta) \to Simp_2(A) - V(\Delta).$$

Notice that if we put $t_1 = t_4 = 0$, then $f = \Delta$. See the Example (3.7)

3.4.2   *Completions of $Simp_n(A)$*

In the example above it is easy to see that elements of the complement of $U(n)$ in the affine scheme $Spec(C(n))$ will be represented by indecomposable, or decomposable representations. A decomposable representation W will, however, not in general be deformable into a simple representation, since good deformations must conserve $End_A(W)$. Therefore, even though we have termed $Spec(C(n))$ a compactification of $Simp_n(A)$, it is a bad *completion*. The missing points *at infinity* of $Simp_n(A)$, should be represented as indecomposable representations, with $End_A(W) = k$. Any such is an iterated extension of simple representations $\{V_i\}_{i=1,2,...s}$, with representation graph Γ (corresponding to an *extension type*, see [18]), and $\sum_{i=1}^{s} dim(V_i) = n$. To simplify the notations we shall write, $|\Gamma| := \{V_i\}_{i=1,2,...s}$. In [16], we treat the problem of classifying all such indecomposable representations, up to isomorphisms. Let us recall the main ideas.

Assume given a family of modules $\{V_i\}$ such that all $Ext_A^1(V_i, V_j)$ are finite dimensional as k-vector spaces. Let Γ be an ordered graph with set of nodes $|\Gamma| = \{V_i\}$. Starting with a first node of Γ, we can construct, in many ways, an extension of the corresponding module V_{i_1} with the module V_{i_2} corresponding to the end point of the first arrow of Γ, then continue, choosing an extension of the result with the module corresponding to the endpoint of the second arrow of Γ, etc. untill we have reached the endpoint of the last arrow. Any finite length module can be made in this way, the oppositely ordered Γ corresponding to a decomposition of the module into simple constituencies, by peeling off one simple sub-module at a time, i.e. by picking one simple sub-module and forming the quotient, picking a second simple sub-module of the quotient and taking the quotient, and repeating the procedure until it stops.

The *ordered* k-algebra $k[\Gamma]$ of the ordered graph Γ is the quotient algebra of the usual algebra of the graph Γ by the ideal generated by all admissible words which are not intervals of the ordered graph. Say $...\gamma_{i,j}(n-1)\gamma_{j,j}(n)\gamma_{j,k}(n+1)...$ is is an interval of the ordered graph, then $\gamma_{i,j}(n-1).\gamma_{j,k}(n+1) = 0$ in $k[\Gamma]$.

Now, let $H(|\Gamma|)$ be the formal moduli of the family $|\Gamma|$. We show in [18], see Proposition 2. above, that any iterated extension of the $\{V_i\}_{i=1}^r$ with *extension type*, i.e. graph, Γ corresponds to a morphism in $\underline{a}_r$,

$$\alpha : H \longrightarrow k[\Gamma].$$

Moreover the set of isomorphism classes of such modules is parametrized by a quotient space of the affine scheme,

$$\underline{A}(\Gamma) := Mor_{\underline{a}_r}(H(|\Gamma|), k[\Gamma]).$$

Let $\alpha \in \underline{A}(\Gamma)$, and let $V(\alpha)$ denote the corresponding iterated extension module, then the tangent space of $\underline{A}(\Gamma)$ at α is,

$$T_{\underline{A}(\Gamma),\alpha} := Der_k(H(|\Gamma|), k[\Gamma]_\alpha),$$

where $k[\Gamma]_\alpha$ is $k[\Gamma]$ considered as a $H(|\Gamma|)$-bimodule via α. The obstruction space for the deformation functor of α is $HH^2(H(|\Gamma|), k[\Gamma])$, and we may, as is explained in [14], [15], compute the complete local ring of $\underline{A}(\Gamma)$ at α. In particular we may decide whether the point is a smooth point of $\underline{A}(\Gamma)$, or not.

The automorphism group G of $k[\Gamma]$, considered as an object of $\underline{a}_r$, has a Lie algebra which we shall call $\mathfrak{g}$. Obviously we have,

$$\mathfrak{g} = Der_k(k[\Gamma], k[\Gamma]).$$

Clearly an iterated extension α with graph Γ will be isomorphic as A-module to $g(\alpha)$, for any $g \in G$. In particular, if $\delta \in \mathfrak{g}$, then $exp(\delta)(\alpha)$ is isomorphic to α as an iterated extension of A-modules, with the same graph as α.

Consider the map,

$$\alpha^* : Der_k(k[\Gamma], k[\Gamma]) \to Der_k(H(\Gamma), k[\Gamma]_\alpha).$$

The image of α^* is the subspace of the tangent space of $\underline{A}(\Gamma)$ at α along which the corresponding module has constant isomorphism class.

Notice that if α is a smooth point, and α^* is not surjectiv then there is a positive-dimensional moduli space of iterated extension modules with graph Γ through α.

The kernel of α^* is contained in the Lie algebra of automorphisms of the module $V(\alpha)$, and should be contained in $End_A(V(\alpha))$. From this follows that if $V(\alpha)$ is indecomposable then $ker\alpha^* = 0$. The Euler type derivations, defined by,

$$\delta_E(\gamma_{i,j}) = \rho_{i,j}\gamma_{i,j}, \ \rho_{i,j} \in k$$

are the easiest to check! Notice however, that there may be discrete automorphisms in G, not of exponential type, leaving α invariant. Notice also that an indecomposable module may have an endomorphism ring which is a non-trivial local ring.

Assume now that we have identified the *non-commutative* scheme of indecomposable Γ-representation, call it $Ind_\Gamma(A)$. Put $Simp_\Gamma(A) :=$

$Simp_n(A) \cup Ind_\Gamma(A)$. Now, repeat the basics of the construction of $Spec(C(n))$ above. Consider for every open affine subscheme $D(s) \subset Simp_\Gamma(A)$, the natural morphism,

$$A \to \varprojlim_{\underline{c} \subset D(s)} O(\underline{c}, \pi)$$

$\underline{c}$ running through all finite subsets of $D(s)$. Put $B_s(\Gamma) := \prod_{V \in D(s)} H^{A(n)}(V)^{com}$, and consider the homomorphism,

$$A \to A(n) \to \prod_{V \in D(s)} H^{A(n)}(V)^{com} \otimes_k End_k(V) \simeq M_n(B_s(\Gamma)).$$

Let $x_i \in A, i = 1, ..., d$ be generators of A, and consider the images $(x_{p,q}^i) \in B_s(n) \otimes_k End_k(k^n)$ of x_i via the homomorphism of k-algebras,

$$A \to B_s(\Gamma) \otimes M_n(k),$$

obtained by choosing bases in all $V \in Simp_\Gamma(A)$. Notice that since V no longer is (necessarily) simple, we do not know that this map is topologically surjectiv.

Now, $B_s(\Gamma)$ is commutative, so the k-sub-algebra $C_s(\Gamma) \subset B_s(\Gamma)$ generated by the elements $\{x_{p,q}^i\}_{i=1,..,d;\ p,q=1,..,n}$ is commutative. We have a morphism,

$$I_s(\Gamma) : A \to C_s(\Gamma) \otimes_k M_n(k) = M_n(C_s(\Gamma)).$$

Moreover, these $C_s(\Gamma)$ define a presheaf, $\mathbf{C}(\Gamma)$, on the Jacobson topology of $Simp_\Gamma(A)$. The rank n free $C_s(\Gamma)$-modules with the A-actions given by $I_s(\Gamma)$, glue together to form a locally free $\mathbf{C}(\Gamma)$-Module $\mathbf{E}(\Gamma)$ on $Simp_\Gamma(A)$, and the morphisms $I_s(\Gamma)$ induce a morphism of algebras,

$$I(\Gamma) : A \to End_{\mathbf{C}(\Gamma)}(\mathbf{E}(\Gamma)).$$

As for every $V \in Simp_\Gamma(A)$, $End_A(V) = k$, the commutator of A in $H^A(V)^{com} \otimes_k End_k(V)$ is $H^A(V)^{com}$. The morphism,

$$\zeta(V) : H^A(V)^{com} \to HH^0(A, H^A(V)^{com} \otimes_k End_k(V))$$

is therefore an isomorphism, and we may assume that the corresponding morphism,

$$\zeta : \mathbf{C}(\Gamma) \to HH^0(A, End_{\mathbf{C}(\Gamma)}(\mathbf{E}(\Gamma)))$$

is an isomorphism of sheaves.

For all $V \in D(s) \subset Simp_\Gamma(A)$ there is a natural projection,

$$\kappa := \kappa(\Gamma) : C_s(\Gamma) \otimes_k M_n(k) \to H^{A(n)}(V)^{com} \otimes_k End_k(V) \simeq M_n(H^{A(n)}(V)^{com}),$$

which, composed with $I_s(\Gamma)$ is the natural homomorphism,

$$A \longrightarrow H^{A(n)}(V)^{com} \otimes_k End_k(V).$$

κ defines a set theoretical map,

$$t : Simp_\Gamma(A) \longrightarrow Spec(\mathbf{C}(\Gamma)),$$

and a natural surjective homomorphism,

$$\hat{\mathbf{C}}(\Gamma)_{t(V)} \to H^{A(n)}(V)^{com}.$$

Categorical properties implies, as usual, that there is another natural morphism,

$$\iota : H^{A(n)}(V) \to \hat{\mathbf{C}}(\Gamma)_{t(V)},$$

which composed with the former is the obvious surjection, and such that the induced composition,

$$A \longrightarrow H^{A(n)}(V)^{com} \otimes_k End_k(V) \to \hat{\mathbf{C}}(\Gamma)_{t(V)} \otimes_k End_k(V),$$

is $I(\Gamma)$ formalized at $t(V)$. From this, and from the definition of $\mathbf{C}(\Gamma)$, it follows that ι is surjective, such that for every $V \in Simp_\Gamma(A)$ there is an isomorphism $H^{A(n)}(V)^{com} \simeq \hat{\mathbf{C}}(\Gamma)_{t(V)}$.

This implies that, formally at a point $V \in Simp_\Gamma(A)$, the local, commutative structure of $Simp_\Gamma(A)$ (as A or $A(n)$-module), and the corresponding local structure of $Spec(\mathbf{C}(\Gamma))$ at V, coincide. We have actually proved the following,

Theorem 3.4.9. *The topological space $Simp_\Gamma(A)$, with the Jacobson topology, together with the sheaf of commutative k-algebras $\mathbf{C}(\Gamma)$ defines a scheme structure on $Simp_\Gamma(A)$, containing an open subscheme, étale over $Simp_n(A)$. Moreover, there is a morphism,*

$$\pi(\Gamma) : Simp_\Gamma(A) \to Spec(ZA(n)),$$

extending the natural morphism,

$$\pi_0 : Simp_n(A) \to Spec(ZA(n)).$$

Proof. As in Theorem (3.4.8) we prove that if $v = t(V), with\ V \in Simp_n(A) \subseteq Simp_\Gamma(A)$, then there exists an open subscheme of $Spec(\mathbf{C}(\Gamma))$ containing only simple modules of dimension n. If v is indecomposables with $End_A(V) = k$ we may look at the homomorphism of $\mathbf{C}(\Gamma)$-modules,

$$End_A(\mathbf{C}(\Gamma)) \otimes End_k(V) \longrightarrow End_A(V) = k.$$

Clearly there is an open neighborhood of v in $Spec(\mathbf{C}(\Gamma))$ containing only indecomposables of dimension n. $\qquad\square$

These morphisms $\pi(\Gamma)$ are our candidates for the possibly different completions of $Simp_n(A)$. Notice that for $W \in Spec(C(n)) - U(n)$, the formal moduli $H^A(W)$ is not always pro-representing. If W is semi-simple, but not simple then $End_A(W) \neq k$. The corresponding modular substratum will, locally, correspond to the semi-simple deformations of W, thus to a closed subscheme of $Spec(C(n)) - U(n) \subset Spec(C(n))$. This follows from the fact that the substratum of modular deformations of any semisimple (but not simple) module V will have a tangent space equal to the invariant space of the action of the $End_k(V)$ on $Ext^1_A(V, V)$, which must be the sum of the tangent spaces of the deformation spaces of the simple components of V.

As we have already remarked, $Spec(C(n))$ is, in a sense, a compactification of $U(n)$. It is, however not the correct *completion* of $U(n)$. In fact, the points of $Spec(C(n)) - U(n)$ may correspond to semi-simple modules, which do not deform into simple n-dimensional modules. We shall in the last chapter return to the study of the (notion of) completion, in connection with the process of *decay* and *creation* of *particles*. Decay occur, at *infinity* in $Simp_n(A)$, see the Introduction.

3.5 Morphisms, Hilbert Schemes, Fields and Strings

Above we have studied moduli spaces of representations of finitely generated k-algebras. We might as well have studied the Hilbert functor, $\mathbf{H}_{A^r}$, of subschemes of length r of the spectrum of the algebra A, or the moduli space $\mathbf{F}(A; R)$, of morphisms, $\kappa : A \to R$, for fixed algebras, A and R. The difference is that whereas for finite n, the set $Simp_n(A)$ has a nice, finite dimensional scheme structure, this is, in general, no longer true for the set, $\mathbf{H}^r_A$ nor for the set of *fields*, $\mathbf{F}(A; R)$, as the physicists call it, unless we put some extra conditions on the fields, so called decorations. If R is Artinian of length n, then the corresponding $\mathbf{F}(A; R)$ does exist and has a nice structure, both as commutative and as non-commutative scheme. The toy model of relativity theory, referred to in the introduction, is in fact modeled on $\mathbf{M}(k[x_1, x_2, x_3], k^2)$, i.e. on the set of surjective homomorphisms $k[x_1, x_2, x_3] \to R = k^2$. And, in all generality, the *space* $\mathbf{F}(A; R)$ has a tangent structure. I fact, depending on the point of view, the tangent space of a morphism $\phi : A \to R$ is equal to,

$$T_{\mathbf{F}(A;R),\phi} = Der_k(A, R)/Triv,$$

where $Triv$ either is 0 or the inner derivations induced by R. Even though there is no obvious algebraic structure on $\mathbf{F}(A; R)$ this general situation is important. It is the basis of our treatment of *Quantum Field Theory*, as we shall see, in the next chapter. There it will be treated in combination with the notion of *clock*.

We may also consider the notion of *string*, in the same language as above. Let us, for the fun of it, make the following :

Definition 3.5.1. A *general string, or a g-string*, is an algebra R together with a pair of Ph-points, i.e. a pair of homomorphisms $\epsilon_i : Ph(R) \to k(p_i)$, corresponding to two points $k(p_i) \in Simp_1(R)$ each outfitted with a tangent ξ_i.

We might have considered any two points $k(p_i) \in Simp_n(Ph(R))$, but since the main properties of the g-strings will be equally well understood restricting to the case $n = 1$, we shall postpone this generalization. For any g-string, consider the *non-commutative tangent space* of the the pair of points,

$$T(R, p_1, p_2) := Ext^1_{PhR}(p_1, p_2).$$

We shall call it the *space of tensions*, between the two points of the string. Consider the space $String_g(A)$ of $g-strings$ in A, i.e. the space of isomorphism classes of algebra homomorphisms $\kappa : A \to R$ where R is a g-string, and where isomorphisms should correspond to isomorphisms of the g-string, thus conserving the two PhR-points. Any g-string in A, $\kappa : A \to R$, induces a unique commutative diagram of algebras, The von Neumann condition imposed on a string κ, is now the following,

$$\epsilon_i \circ Ph\kappa \circ d = \kappa_* \xi_i =: \overline{\xi}_i = 0, \ i = 1 \vee i = 2,$$

which, if $x_j, j = 1, .., n$ and $\sigma_l, l = 1, .., p$ are parameters of A respectively R, is equivalent to the condition,

$$\frac{\partial x_j}{\partial \sigma_l}(p_i) = 0, j = 1, ..., n, \ l = 1, .., p, \ i = 1 \vee 2.$$

Notice also that, since any derivation $\xi \in Der_k(A, R)$ has a natural lifting to a derivation $\overline{\xi} \in Der_k(PhA, PhR)$ defined by simply putting $\overline{\xi}(a) = d(\xi(a))$, we find, using the general machinery of deformations of diagrams, see [14], that any family of morphisms κ induces a family of the above diagram. If $\tau_k, \ k = 1, ..., d$ are parameters of such a family, $\mathbf{M} = Spec(M)$, then $d\tau_i \in PhM$ corresponds to a derivation,

$\tau_i \in Der_k(A, R)$, and therefore to tangents $\overline{\xi}_i$, $i = 1, 2$, of $Simp_1(A)$ at the two points $k(p_i)$. The Dirichlet condition on the string is now,

$$\overline{\xi}_i = 0, \ i = 1 \vee 2,$$

which is equivalent to the condition,

$$\frac{\partial x_j}{\partial \tau_l}(p_i) = 0, j = 1, ..., n, \ l = 1, .., p, \ i = 1 \vee 2.$$

These conditions will define new moduli spaces which we shall call $String_R^{vN}(A)$ and $String_R^D(A)$, respectively. In the affine case the structure of these spaces is a problem, however we may of course do everything above for A and R replaced by projective schemes, and then all the moduli spaces exist as classical schemes. The volume form of the space the string is fanning out will give us a an action functional, S, defined on $String_R(A)$, see next chapter.

Let us end this sketch by noticing that there is a non-commutative deformation theory for fields, just as there is one for representations of associative algebras. In fact, let $\{\kappa_i\}_{i=1,...,r}$ be a finite family of fields, and consider for every pair $(i, j) | 1 \leq i, j \leq r$ the A-bimodule $R_{i,j}$ where κ_i defines the left module structure, and κ_j the right hand structure. Then copying the definition for the non-commutative deformation functor for representations, replacing $Hom_k(V_i, V_j)$ by $R_{i,j}$, we may prove most of the results referred to at the beginning of this chapter. This may be of interest in relation with the problems of *interactions* treated in the last chapter of this book.

Example 3.4. (i) Let us go back to Example (1.1)(ii). It follows that the string of dimension 0, $R = k^2$, $Ph(R) = k < x, dx > /((x^2 - r^2), (xdx + dxx))$, has unique points, $k(\pm r)$. The space of tensions is of dimension 1, the von Neumann condition is automatically satisfied, and the moduli space of k^2-strings in $A = k[x_1, x_2, x_3]$ is nothing but $\underline{H} := Speck[t_1, ..., t_6] - \underline{\Delta}$. If we consider the string with $R = k[x]/(x^2)$, $PhR = k[x, dx]/(x^2, (xdx + dxx))$, then we see that there is just one point of R, but a line of point for PhR, all corresponding to $x = 0$ in R. Therefore there is a 2-dimensional space of strings with the same R. Compare this with the blow-up $\tilde{\underline{H}}$, see [20]. (ii) In dimension 1 the simplest *closed string* is given by, $R = k[x, y]/(f)$, with $f = x^2 + y^2 - r^2$, such that $PhR = k < x, y, dx, dy > /(f, [x, y], d[x, y], df)$, and with the two points, $\epsilon_i : PhR \rightarrow k(p_i)$, defined by the actions on $k(p_i) := k$, given by, $x_i, y_i, (dx)_i, (dy)_i$, $i = 1, 2$. It is easy to see that the vectors, $\xi_i := ((dx)_i, (dy)_i)$ are tangent vectors to the circle at the points p_i, and if $p_1 \neq p_2$ we find that $Ext^1_{PhR}(k(p_1), k(p_2)) = k$. The von Neumann condition is, $\xi_i = 0$, $i = 1 \lor i = 2$, and this clearly means that $\frac{\partial x}{\partial \sigma} = \frac{\partial y}{\partial \sigma} = 0$ at one of the points p_i. The 1-dimensional *open string* is now left as an exercise.

Chapter 4

Geometry of Time-spaces and the General Dynamical Law

Given a finitely generated k-algebra, and a natural number n, we have in Chapter 3 constructed a scheme $Simp_n(A)$, and a versal family,

$$\tilde{\rho} : A \to End_{U(n)}(\tilde{V})$$

defined on an étale covering $U(n)$ of $Simp_n(A)$. $U(n)$ is an open subscheme of an affine scheme $Spec(C(n))$, and the versal family is, in fact, defined on $C(n)$.

4.1 Dynamical Structures

We would now like to use this theory for the k-algebra $Ph^\infty(A)$ of Chapter 2. However, $Ph^\infty(A)$ is rarely of finite type. We shall therefore intoduce the notion of *dynamical structure*, and the *order* of a dynamical structure, to reduce the problem to a situation we can handle. This is also what physicists do, they invoke a parsimony principle, or an action principle, originally proposed by Fermat, and later by Maupertuis, with exactly this purpose, reducing the preparation needed to be able to see ahead, see Chapter 2.

Definition 4.1.1. A *dynamical structure*, σ, is a two-sided δ-stable ideal $(\sigma) \subset Ph^\infty(A)$, such that

$$\mathbf{A}(\sigma) := Ph^\infty(A)/(\sigma),$$

the corresponding, *dynamical system*, is of finite type. A dynamical structure, or system, is *of order n* if the canonical morphism,

$$\sigma : Ph^{(n-1)}(A) \to A(\sigma)$$

is surjective. If A is generated by the *coordinate functions*, $\{t_i\}_{i=1,2,...,d}$ a dynamical system of order n may be defined by a *force law*, i.e. by a system of equations,

$$\delta^n t_p = \Gamma^p(\underline{t}_i, \underline{dt}_j, \underline{d}^2 t_k, .., \underline{d}^{n-1} t_l), \ p = 1, 2, ..., d.$$

Put,

$$\mathbf{A}(\sigma) := Ph^\infty(A)/(\delta^n t_p - \Gamma^p)$$

where $\sigma := (\delta^n t_p - \Gamma^p)$ is the two-sided δ-ideal generated by the defining equations of σ. Obviously δ induces a derivation $\delta_\sigma \in Der_k(\mathbf{A}(\sigma), \mathbf{A}(\sigma))$, also called the Dirac derivation, and usually just denoted δ.

Notice that if σ_i, $i = 1, 2$, are two different order n dynamical systems, then we may well have,

$$\mathbf{A}(\sigma_1) \simeq \mathbf{A}(\sigma_2) \simeq Ph^{(n-1)}(A)/(\sigma_*),$$

as k-algebras, see the Introduction.

4.2   Quantum Fields and Dynamics

For any integer $n \geq 1$ consider the schemes $Simp_n(\mathbf{A}(\sigma))$ and $Spec(C(n))$, and the corresponding (almost uni-) versal family,

$$\tilde{\rho} : \mathbf{A}(\sigma)) \to End_{Spec(C(n))}(\tilde{V}) \simeq M_n(C(n)).$$

The Dirac derivation $\delta \in Der_k(\mathbf{A}(\sigma), \mathbf{A}(\sigma))$, composed with ρ, corresponding to any element $v \in Simp_n(\mathbf{A}(\sigma))$, defines, as explained in Chapter 2, a tangent vector of $Simp_n(\mathbf{A}(\sigma))$ at the point v, thus a distribution on $Simp_n(\mathbf{A}(\sigma))$. The reason why the Dirac derivation, does not define a unique vector-field is, of course, that the structure morphisms of the simple modules can be scaled by any non-zero element of the field k. However, once we have chosen a versal family for the moduli space $Simp_n(A(\sigma))$, defined on $Spec(C(n))$, the Dirac derivation δ induces, by composition with $\tilde{\rho}$, an element,

$$\tilde{\delta} \in Der_k(A(\sigma), End_{C(n)}(\tilde{V})).$$

which obviously induces a well defined vector field $\xi \in \Theta_{U(n)}$, in the distribution defined by δ. Now, to any vectorfield η of $Spec(C(n))$, i.e. for any derivation $\eta \in Der_k(C(n))$, there is a unique element,

$$\eta' \in Der_k(A(\sigma), End_{C(n)}(\tilde{V})),$$

defined by,

$$\eta'(a) = \eta(\tilde{\rho}(a)),$$

where we have identified $\tilde{\rho}(a)$ with an element of $M_n(C(n))$. Suppose there exist a (rational) derivation $[\delta] \in Der_k(C(n))$, lifting the vector field ξ in $Simp_n(\mathbf{A}(\sigma))$, defined by the Dirac derivation, then by construction of $C(n)$, and of the versal family $\tilde{\rho}$, in general, we find that $\tilde{\delta} - [\delta]'$ is an inner derivation, defined by some $Q \in End_{C(n)}(\tilde{V})$.

This is the situation which we shall find ourselves in, in the sequel, see the Examples (4.1) and (4.3).

In general, we have the fundamental result:

Theorem 4.2.1. *Formally, at any point $v \in U(n)$, with local ring $\hat{C(n)}_v$, there is a derivation $[\delta] \in Der_k(\hat{C(n)}_v)$, and an Hamiltonian $Q \in End_{\hat{C(n)}_v}(\tilde{V}_v)$, such that, as operators on $\tilde{V}_v$, we must have,*

$$\delta = [\delta] + [Q, -].$$

This means that for every $a \in \mathbf{A}(\sigma)$, considered as an element $\tilde{\rho}(a) \in M_n(\hat{C(n)}_v)$, $\delta(a)$ acts on $\tilde{V}_v$ as

$$\tilde{\rho}(\delta(a)) = \xi(\tilde{\rho}(a)) + [Q, \tilde{\rho}(a)].$$

Proof. Suppose the family,

$$\tilde{\rho} : \mathbf{A}(\sigma) \to End_{Spec(C(n))}(\tilde{V}) \simeq M_n(C(n)).$$

had been the universal family of a fine moduli space. Then any (infinitesimal) automorphism of $\mathbf{A}(\sigma)$ would have been squared off by an (infinitesimal) automorphism of $End_{Spec(C(n))}(\tilde{V}) = M_n(C(n))$. Given a derivation δ of $\mathbf{A}(\sigma)$ there is an induced homomorphism, $\mathbf{A}(\sigma) \to \mathbf{A}(\sigma) \otimes_k k[\epsilon]$. Composed with the natural homomorphism, $\mathbf{A}(\sigma) \otimes_k k[\epsilon] \to C(n) \otimes_k k[\epsilon] \otimes End_k(V) \simeq M_n(C(n) \otimes_k k[\epsilon])$, we find a lifting of the family, and we know that there should exist a morphism, $C(n) \to C(n) \otimes_k k[\epsilon]$ defining an isomorphic lifting. Now, this induces a derivation $[\delta]$, of $C(n)$ plus a trivial derivation; $ad(Q)$, of $M_n(C(n))$, exactly as we want. Recall also that any derivation of $M_n(C(n))$ is a sum of a derivation of $C(n)$ plus an inner derivation $ad(Q), Q \in M_n(C(n))$. Since our space $U(n) \subset Spec(C(n))$ is an étale covering of the moduli space $Simp_n(\mathbf{A}(\sigma))$, and since our versal family is only defined over $U(n)$, we need to restrict to the formal case. Since $\hat{C(n)}_v \simeq \hat{O}_{Simp_n(\mathbf{A}(\sigma)),v}$ this case is clear by the general deformation theory, just like above. $\qquad\square$

As pointed out above, in the examples that we shall meet in the sequel, there are local (or even global) extensions of this result, where $[\delta]$ and Q may be assumed to be defined (rationally) on $C(n)$.

This Q, the *Hamiltonian* of the system, is in the singular case, when $[\delta] = 0$, also called the Dirac operator, and sometimes denoted δ-slashed, see e.g. [29], or other texts on Connes' spectral tripples. In fact, a *spectral tripple* is composed of a vector space like $\tilde{V}$, together with a Dirac operator, like Q, and a complexification etc.

If $[\delta] = 0$, it is also easy to see that what we have observed implies that Heisenberg's and Schrödinger's way of doing quantum mechanics, are strictly equivalent.

In line with our general philosophy, we shall consider ξ, or $[\delta]$ as measuring *time* in $Simp_n(\mathbf{A}(\sigma))$, respectively in $Spec(C(n))$.

Assume for a while that $k = \mathbf{R}$, the real numbers, and that our constructions go through, as if k were algebraically closed. Let $v(\tau_0) \in Simp_n(\mathbf{A}(\sigma))$ be an element, an *event*. Suppose there exist an integral curve γ of ξ through $v(\tau_0) \in Simp_1(C(n))$, ending at $v(\tau_1) \in Simp_1(C(n))$, given by the automorphisms $e(\tau) := exp(\tau\xi)$, for $\tau \in [\tau_0, \tau_1] \subset \mathbf{R}$. The supremum of τ for which the corresponding point, $v(\tau)$, of γ is in $Simp_n(\mathbf{A}(\sigma))$ should be called the *lifetime* of the particle. We shall see that it is relatively easy to compute these lifetimes, when the fundamental vector field ξ has been computed.

This, however, is certainly not so easy, see the examples (3.4)-(3,8).

Let now $\psi(\tau_0) \in \tilde{V}(v_0) \simeq V$ be a (classically considered) state of our *quantum system*, at the time τ_0, and consider the (uni)versal family,

$$\tilde{\rho} : \mathbf{A}(\sigma) \longrightarrow End_{C(n)}(\tilde{V})$$

restricted to $U(n) \subseteq Simp_1(C(n))$, the étale covering of $Simp_n(\mathbf{A}(\sigma))$. We shall consider $\mathbf{A}(\sigma)$ as our *ring of observables*.

What happens to $\psi(\tau_0) \in V(0)$ when *time* passes from τ_0 to τ, along γ? This is obviously a question that has to do with whether we choose to consider the Heisenberg or the Schrödinger picture. In fact, if we consider the formal flow $exp(t\delta)$ defined on the ring of observables, then putting,

$$u(\tau) := exp(\tau\nabla_\xi),$$

where,

$$\nabla_\xi := \xi + Q \in End_k(\tilde{V}),$$

is a connection on $\tilde{V}$, we obtain for every $\psi \in \tilde{V}$, and every $a \in A(\sigma)$, that the equation,

$$u(\tau)(\tilde{\rho}(exp(-\tau\delta)(a))(\psi)) = \tilde{\rho}(a)(u(\tau)(\psi))$$

holds formally, at least up to first order. In fact, up to order one, in τ, the left hand side is equal to

$$\tilde{\rho}(a)(\psi) - \tau\tilde{\rho}(\delta(a))(\psi) + \tau\xi(\tilde{\rho}(a)(\psi)) + \tau Q\tilde{\rho}(a)(\psi),$$

and the right hand side is,

$$\tilde{\rho}(a)(\psi) + \tau\tilde{\rho}(a)(\xi(\psi)) + \tau\tilde{\rho}(a)(Q(\psi)).$$

Noticing that $\xi(\tilde{\rho}(a)(\psi)) = \xi(\tilde{\rho}(a))(\psi) + \tilde{\rho}(a)(\xi(\psi))$, and using (4.2.1.) we find that the two sides are equal.

This means that $exp(\tau\delta)$ keeps $\tilde{V}$ fixed within its conjugate class, up to first order in τ. Thus, an element $\psi \in \tilde{V}$ which is *flat* with respect to the connection ∇_ξ, above γ, has the property that,

$$\tilde{\rho}(\delta(a))\psi = \nabla_\xi(\tilde{\rho}(a)(\psi)),$$

for all $a \in A(\sigma)$.

It is therefore reasonable to consider any flat state, $\psi(t) \in \tilde{V}$, as the time development of $\psi(0) \in V(0)$. Clearly, the flat states $\psi \in \tilde{V}$, are solutions of the differential equation,

$$\xi(\psi) = -Q(\psi), i.e. \ \frac{\partial \psi}{\partial \tau} = -Q(\psi).$$

which, if we accept that time is the parameter τ of the integral curve γ, is the Schrödinger equation.

Notice that, in the classical quantum-theoretical case, one works with one fixed representation, corresponding to what we have called a singular point of ξ. This implies that we are looking at a representation V with $\xi(v) = 0$, and so we have *no time*. What we call time is then the parameter of the one-parameter automorphism group $u(\tau) := exp(\tau Q)$ acting on V. This also leads to a Schrödinger equation, and to the next result, proving that ψ is completely determined, along any integral curve γ by the value of $\psi(\tau_0)$, for any $\tau_0 \in \gamma$.

Theorem 4.2.2. *The evolution operator* $u(\tau_0, \tau_1)$ *that changes the state* $\psi(\tau_0) \in \tilde{V}(v_0)$ *into the state* $\psi(\tau_1) \in \tilde{V}(v_1)$, *where* $\tau_1 - \tau_0$ *is the length of the integral curve* γ *connecting the two points* v_0 *and* v_1, *i.e. the time passed, is given by,*

$$\psi(\tau_1) = u(\tau_0, \tau_1)(\psi(\tau_0)) = exp[\int_\gamma Q(\tau)d\tau] \ (\psi(\tau_0)),$$

where $exp \int_\gamma$ *is the non-commutative version of the ordinary action integral, essentially defined by the equation,*

$$exp[\int_\gamma Q(\tau)dt] = exp[\int_{\gamma_2} Q(\tau)d\tau] \circ exp[\int_{\gamma_1} Q(\tau)d\tau]$$

where γ *is* γ_1 *followed by* γ_2.

Proof. This is a well known consequence of the Schrödinger equation above. In classical quantum theory one uses a *chronological operator* τ, to keep track of the *intermediate* time-steps that, in our case, are well defined by the integral curve γ, the existence of which we assume. The formula above is related to what the physicists call the Dyson series, see [30], Vol I, Chap. 9, or [2], Chapitre 6. Since we have given the real curve γ parametrized by τ we may look at γ as a closed interval of **R**, $I := [0, \tau]$. Subdivide I into m equal intervals, $[i\Delta\tau, (i+1)\Delta\tau]$, and see that the Schrödinger equation gives, formally,

$$\psi((i+1)\Delta\tau) = exp(\Delta\tau Q)(\psi(i\Delta\tau)).$$

Writing out the power series in $\Delta\tau$, and summing up we find the formula above.   □

The problem of integrating the differential equations above, i.e. finding algebraic geometric formulas for the integral curves of $\xi = [\delta]$, is a classical problem, and we may use a technique already well known to Hamilton and Jacobi. In fact, assuming that $A = k[t_1, ..., t_n]$, and that σ is determined by the following *force-laws*,

$$d^2 t_i = \Gamma^i(t_1, ..., t_n, dt_1, ..., dt_n)$$

we have that,

$$\mathbf{A}(\sigma) = Ph^\infty(A)/(\sigma), \quad \delta = \sum_{i=1}^{n} (dt_i \frac{\partial}{\partial t_i} + \Gamma^i \frac{\partial}{\partial dt_i}).$$

We may try to solve the equation,

$$\delta\theta = 0$$

in the ring $\mathbf{A}(\sigma)$. Obviously the set of solutions form a sub-ring of $\mathbf{A}(\sigma)$, the ring of invariants, and we have the following easy result,

Proposition 4.2.3. *(i): Let* $\Theta = ker\delta$, *be the subring of invariants in* $\mathbf{A}(\sigma)$, *and let* $\rho : \mathbf{A}(\sigma) \to End_k(V)$ *be an n-dimensional representation for which the tangent space of* $Simp_n(\mathbf{A}(\sigma))$, *at* V, $Ext^1_{\mathbf{A}(\sigma)}(V, V) = 0$; *or suppose* V *corresponds to a point* $\underline{t} \in Simp_n(\mathbf{A}(\sigma))$ *for which* $\xi(\underline{t}) = 0$, *then any* $\theta \in \Theta$ *is constant in* V, *i.e.* $[Q, \rho(\theta)] = 0$, *so that the eigenvectors of* Q *are eigenvectors for* θ.

(ii): Consider for any n the universal family,

$$\tilde{\rho} : \mathbf{A}(\sigma) \to End_{C(n)}(\tilde{V}).$$

and let $\theta \in \Theta$, then

$$Trace\tilde{\rho}(\theta) \in C(n)$$

is constant along any integral curve of ξ in $Simp_n(\mathbf{A}(\sigma))$, i.e.

$$[\delta](Trace\tilde{\rho}(\theta)) = 0$$

Proof.   (i) Suppose $\delta(\theta) = 0$, and consider the dynamical equation,

$$\delta = [\delta] + [Q, -],$$

where we may assume $[\delta] = \xi$. If the tangent space of V is trivial, then obviously $[\delta] = 0$, therefore $\delta(\theta) = 0$ implies $[Q, \rho(\theta)] = 0$.

(ii) If $\delta(\theta) = 0$, we must have, in $End_{C(n)}(\tilde{V})$,

$$0 = Trace\xi(\tilde{\rho}(\theta)) + Trace[Q, \tilde{\rho}(\theta)] = Trace\xi(\tilde{\rho}(\theta)) = \xi(Trace\tilde{\rho}(\theta)). \quad \square$$

Notice that we find the same formulas if we extend the action of $\mathbf{A}(\sigma)$ to $\tilde{V}_{\mathbf{C}} := \tilde{V} \otimes_{\mathbf{R}} \mathbf{C}$. This is what turns out to be the interesting case in quantum physics. It is easy to see that if $A = k[x_1, ..., x_d] \subset \mathbf{A}(\sigma)$ is a polynomial algebra, and σ is a second order force-law, such as,

$$d^2 x_i = \sum \Gamma^i_{j,k} dx_j dx_k, \ i, j, k = 1, 2, ..., d,$$

then, if we have chosen a versal family,

$$\tilde{\rho} : \mathbf{A}(\sigma) \rightarrow End_{C(n)}(\tilde{V})$$

for the simple n-dimensional representations, we obtain another, complexified, versal family,

$$\tilde{\rho}_{\mathbf{C}} : \mathbf{A}(\sigma) \rightarrow End_{C(n)}(\tilde{V}_{\mathbf{C}})$$

with exactly the same formal properties by defining, $\tilde{\rho}_{\mathbf{C}}(x_i) = \tilde{\rho}(x_i)$, $\tilde{\rho}_{\mathbf{C}}(dx_i) = \imath\tilde{\rho}(dx_i)$, and putting $\xi_{\mathbf{C}} = \imath\xi$, $Q_{\mathbf{C}} = \imath Q$.

A section ψ of the complex bundle $\tilde{V}$, a *state*, is now a function on the moduli space $Simp_n(A(\sigma))$, not a function on the *configuration space*, $Simp_1(A)$, or $Simp_1(A(\sigma))$. The value $\psi(v) \in \tilde{V}(v)$ of ψ, at some point $v \in Simp_n(A)$, will also be called a *state*, at the *event v*.

$End_{C(n)}(\tilde{V})$ induces a complex bundle, of *operators*, on $Simp_n(\mathbf{A}(\sigma))$. A section, ϕ of this bundle will be called a *quantum field*. In particular, any element $a \in (\mathbf{A}(\sigma)$ will , via the versal family map $\tilde{\rho}$, define a quantum field. The set of quantum fields therefore form a natural k-algebra.

Physicists will tend to be uncomfortable with this use of their language. A classical quantum field for any traditional physicist is, usually, a *function* $\psi(p, \sigma, n)$, defined on *configuration space*, (which is not our $Simp_n(\mathbf{A}(\sigma))$) with values, in the polynomial algebra generated by certain *creation* and *annihilation* operators in a *Fock-space*.

As we shall see, this interpretation may be viewed as a special case of our general notion.

4.3   Classical Quantum Theory

Now, assume $A = k[x_1, ..., x_d]$ and that the k-algebra $C(n)$ is generated by $\{t_1, ..., t_s\}$. Let us denote by $\underline{t}$ a point of $Simp_1(C(n))$. Since the *configuration-space* coordinates x_i commute, we may find rational sections

$$|\underline{x}_\nu(\underline{t}) >\in \tilde{V}, \ \nu = 1, ..., n,$$

that are eigenvectors for all x_i, such that any point in $U(n)$ correspond to n points in the *configuration space* given by the n possibilities, $(x_{1,\nu}(\underline{t}), ..., x_{d,\nu}(\underline{t})) \ \nu = 1, ..., n$, where,

$$\tilde{\rho}(x_i)(|\underline{x}_\nu(\underline{t}) >) = x_{i,\nu}(\underline{t})|\underline{x}_\nu(\underline{t}) > .$$

In general, the observables dx_i, i=1,...,d, do not commute, but for every i we can still find eigenvectors,

$$|dx_{i,\nu}(\underline{t}) >\in \tilde{V}_{\mathbf{C}}, \ \nu = 1, ..., n,$$

such that,

$$\tilde{\rho}(dx_i)(|dx_{i,\nu}(\underline{t}) >) = dx_{i,\nu}(\underline{t})|dx_{i,\nu}(\underline{t}) > .$$

This will be treated in the section *Grand picture, Bosons, Fermions, and Supersymmetry*, where we also focus on the notion of *locality of interaction*.

Pick a point $\underline{t}_0 \in \underline{U}(n)$ and let $v_0 \in Simp_n(A(\sigma))$ represent the corresponding simple module, and assume we have computed the integral curve γ parametrized by τ, through v_0, ending at v_1, represented by $\underline{t}_1 \in \underline{U}(n)$. Suppose moreover that we have lifted this curve to $U(n)$, thus beginning in $\underline{t}_0$ and ending in $\underline{t}_1$. The evolution operator $u(\tau_0, \tau_1)$ acts upon each $|\underline{x}_\nu(\underline{t}_0) >, \nu = 1, ..., n$. The result will have the form,

$$u(\tau_0, \tau_1)(|\underline{x}_\nu(\underline{t}_0) >) = \sum_{\mu=1,..,n} \gamma_{\nu,\mu}(\tau)|\underline{x}_\mu(\underline{t}_1) >$$

and,

$$u(\tau_0, \tau_1)(|dx_{i,\nu}(\underline{t}_0) >) = \sum_{\mu=1,..,n} \gamma_{i,\nu,\mu}(\tau)|dx_{i,\mu}(\underline{t}_1) >,$$

where each $\gamma_{\nu,\mu}(\tau)$ and $\gamma_{i,\nu,\mu}(\tau)$ is a kind of action integral maybe related to some classical Lagrangian.

We might now consider the following laboratory situation, in which there are n^3 cells $\{X_{q_1,q_2,q_3}\}_{q_i=1,...,n,i=1,2,3}$, disposed in a structure like a grid of space, with coordinates (q_1, q_2, q_3), and each capable of clicking, if entered by a *particle*. Each cell is outfitted with n^3 sub-cells, $\{Y^{p_1}, Y^{p_2}, Y^{p_3}, p_i = 1, ..., n, \ i = 1, 2, 3\}$, forming a system sub-cells, $\{Y^{p_1,p_2,p_3}_{q_1,q_2,q_3}\}_{p_i,q_i=1,...,n,i=1,2,3}$, each capable of clicking if the particle that enters is outfitted with a certain momentum. Let us talk about these clicks as a q-click and a p-click respectively. Interpreting $\{X_{q_1,q_2,q_3},\ q_1 \leq n_1, q_2 \leq n_2, q_3 \leq n_3, n_1 + n_2 + n_3 = n\}$ as a basis of eigenvectors for the space-observables x_1, x_2, x_3, and considering $\{Y^{p_1}, Y^{p_2}, Y^{p_3}, p_i = 1, ..., n,\ i = 1, 2, 3\}$, as a basis of eigenvectors for the momenta-observables dx_1, dx_2, dx_3, for some versal family of n-dimensional simple representations $\tilde{V}$, defined on the k-algebra $C(n)$. The possible outcomes of a measurement performed at time τ are now limited to hearing a q-click in one of the n^3 points in space, corresponding to the eigenvalues of x_1, x_2, x_3, and for each such q-click, hearing a different p-click corresponding to one of the n^3 eigenvalues of dx_1, dx_2, dx_3. The experiment might consist of letting a beam of particles stream out of an outlet situated at one of the cells, say the one corresponding to the origin $X_{0,0,0}$ of the q-grid. One checks the distribution of p-clicks from the sub-cells $\{Y_{0,0,0}^{p_1,p_2,p_3}\}$, say $\{\beta_{p_1}, \beta_{p_2}, \beta_{p_3}\}$. Now, suppose we have chosen a simple representation $V(\tau_0)$ such that,

$$X_{0,0,0} = \sum_{\underline{p}_i=1}^{n} \beta_{p_i} Y^{p_i},\ i = 1, 2, 3,$$

then we measure time τ along the curve γ of $\underline{U}(n)$, starting at the point corresponding to $V(\tau_0)$, and compute,

$$U(\tau_0, \tau_1)(X_{0,0,0}) = \psi(\tau_1) = \sum_{\underline{q}} \alpha_{q_1,q_2,q_3} X_{q_1,q_2,q_3}.$$

We might then want to interpret the family $|\alpha_{q_1,q_2,q_3}|^2/|\psi(\tau_1)|^2$ as the probability distribution, at time τ_1, for finding the particle, at the corresponding point. And, correspondingly, one would be tempted to consider the normalized squares of the coefficients in the development,

$$X_{q_1,q_2,q_3}(\tau_1) = \sum_{\underline{p}_i=1}^{n} \beta_{q_1,q_2,q_3}^{p_i} Y^{p_i}(\tau_1),\ i = 1, 2, 3,$$

as the probability distribution for momenta observed at the point $\underline{q}(\tau_1)$. However, we have to be careful, we have assumed that we might find an *object* $\tilde{V}$ with the properties corresponding to our *preparation*. This may be possible, as we shall see in an example, see (4.7), but the interpretation of the coefficients α and β as probabilities, will probably depend upon the introduction of Hermitian norms on the representation $\tilde{V}$. Anyway, this seems to lead to a kind of generalized Feynman's path integral. For a good exposition, for mathematicians, of path integrals, see [3].

4.4   Planck's Constant(s) and Fock Space

In [19] we treated the case of a conservative system, i.e. where the vector field ξ or $[\delta]$ in $Simp_n(\mathbf{A}(\sigma))$ is singular, i.e. vanishes, at the point $v \in Simp_n(\mathbf{A}(\sigma))$ corresponding to the representation V, and where therefore the Hamiltonian Q is both the *time* and *the energy operator*, at the same time. See Examples (4.2) and (4.3) where we show how to compute these singularities in some classical cases.

We found, in this situation, see [20], or Chapter 1, that there is a notion of *Planck's constant* $\hbar$, with the ordinary properties.

Let $\{v_i\}_{i \in I}$ be a basis of V (no longer assumed to be finite dimensional), formed by eigenvectors of Q, and let the eigenvalues be given by,

$$Q(v_i) = \kappa_i v_i.$$

Consider the set $\Lambda(\delta)$ of real numbers λ defined by,

$$\Lambda(\delta) := \{\lambda \in \mathbf{R} | \ \exists f_\lambda \in Ph^\infty(A), f_\lambda \neq 0, \rho_V(\delta(f_\lambda)) = \lambda \rho_V(f_\lambda) \in End_k(V)\}.$$

Since $\delta = [Q, -]$ is a derivation, if f_λ and f_μ are eigenvectors for δ in V, then if $f_\lambda f_\mu$ is non-trivial, it is also an eigenvector, with eigenvalue $\lambda + \mu$, implying that if $\lambda, \mu \in \Lambda(\delta)$, with $f_\lambda f_\mu \neq 0$, we must have $\lambda + \mu \in \Lambda(\delta)$. Now,

$$\lambda f_\lambda \cdot v_i = \delta(f_\lambda) \cdot v_i = (Qf_\lambda - f_\lambda Q(v_i) = Q(f_\lambda \cdot v_i) - \kappa_i f_\lambda \cdot v_i,$$

implying,

$$Q(f_\lambda \cdot v_i) = (\kappa_i + \lambda) \cdot (f_\lambda \cdot v_i).$$

If $f_\lambda \cdot v_i \neq 0$, it follows that: $\kappa_i + \lambda = \kappa_j$ for some $j \in I$. Therefore

$$f_\lambda \cdot v_i = \alpha v_j, \ \alpha \in \mathbf{R}, \text{ and } \lambda = \kappa_i - \kappa_j.$$

and so,

$$\Lambda(\delta) \subset \{\kappa_i - \kappa_j | \ i, j\},$$

Planck's constant $\hbar$ should be a generator of the monoid $\Lambda(\delta)$, when this is meaningful.

We can show, see Example (4.1) and (4.3), that for the classical oscillator $\Lambda(\delta)$ is an infinite additive monoid. See also that when $\{f_\lambda\}_\lambda$ generate $End_k(V)$ we must have $\Lambda(\delta) = \{\kappa_i - \kappa_j | \ i, j\}$, and that when $\hbar$ *tends to 0*, any $f \in Ph^\infty(A)$ maps every eigenspace $V(\kappa_i)$ into itself. In the generic case when all κ_i are different, the image of $Ph^\infty(A)$ into $End_k(V)$ becomes commutative, a ring of functions on the spectrum of Q.

The above introduction of Planck's constant(s), also make sense in general, i.e. for the versal family of $Simp(\mathbf{A}(\sigma))$. In fact, since

$$[Q, \tilde{\rho}(-)] = \tilde{\rho}\delta - [\delta]\tilde{\rho} : \mathbf{A}(\sigma) \longrightarrow End_{C(n)}(\tilde{V})$$

is a derivation, we show that the set,

$$\Lambda(\sigma) := \{\lambda \in C(n)|\exists f_\lambda \in \mathbf{A}(\sigma), f_\lambda \neq 0, [Q, \tilde{\rho}(\delta(f_\lambda))] = \lambda\tilde{\rho}(f_\lambda)\},$$

is a generalized additive monoid, i.e. if for $\lambda, \lambda' \in \Lambda(\sigma)$ the product $f_\lambda f_{\lambda'}$ is non-trivial, then $\lambda + \lambda' \in \Lambda(\sigma)$.

Let $\hbar_l \in k$ be generators of $\Lambda(\delta)$. These are our Planck's constants, see examples (3.7) and (3.8). Now, assume there exists a $C(n)$-module basis $\{\tilde{\psi}_i\}_{i \in I}$ of sections of $\tilde{V} = C(n) \otimes V$, formed by eigenfunctions for the Hamiltonian, i.e. such that

$$Q(\tilde{\psi}_i) = \kappa_i \tilde{\psi}_i, \ i \in I,$$

where $\kappa_i \in C(n)$. An element such as $\tilde{\psi}_i \in \tilde{V}$ is usually considered as a *pure state*, with *energy* $\kappa_i \in C(n)$, depending on time, i.e. depending on τ, the length along the integral curve γ. It is also considered as as an *elementary particle* (since $\tilde{V}$ is, by assumption, simple). As in §1 we find,

$$\lambda\tilde{\rho}(f_\lambda)(\tilde{\psi}_i) = Q(\tilde{\rho}(f_\lambda)(\tilde{\psi}_i)) - \tilde{\rho}(f_\lambda)(Q(\tilde{\psi}_i))$$
$$= Q(\tilde{\rho}(f_\lambda)(\tilde{\psi}_i)) - \kappa_i \tilde{\rho}(f_\lambda)(\tilde{\psi}_i)$$

implying,

$$Q(\tilde{\rho}(f_\lambda)(\tilde{\psi}_i)) = (\kappa_i + \lambda)\tilde{\rho}(f_\lambda)(\tilde{\psi}_i).$$

By assumption, if $\tilde{\rho}(f_\lambda)(\tilde{\psi}_i) \neq 0$ it must be an eigenvector of Q, with eigenvalue, say $\kappa_j = \kappa_i + \lambda$. It follows that we have,

$$\Lambda(\sigma) \subset \{\kappa_j - \kappa_i| \ i, j \in I\}.$$

To prove that the two sets are equal we need some extra conditions on the nature of $\mathbf{A}(\sigma)$ and $\tilde{\rho}$. If $\{\tilde{\rho}(f_\lambda)\}_\lambda$ generate $End_{C(n)}(\tilde{V})$, then the equality must hold, since then $\{\tilde{\rho}(f_\lambda)(\psi(0))\}_\lambda$ must generate $\tilde{V}$ as $C(n)$-module, and therefore contain multiples of all ψ_j, so that any κ_l must be equal to $\kappa_0 + \lambda$ for some λ.

Notice that if $\hbar$ *goes to* 0, meaning that $[Q, \tilde{\rho}(a)] = 0$, for all $a \in \mathbf{A}(\sigma)$, then all $a \in \mathbf{A}(\sigma)$ must commute with Q, and so $A(\sigma)$ acts diagonally on the spectrum of Q.

Notice also that if, at a point $v \in \gamma$, $\hbar(v) \neq 0$ as an element of $k = \mathbf{R}$, it is clearly reasonable to redefine δ and $Q(v)$ by dividing both with $\hbar(v)$. Then the *energy differences* of $(1/\hbar(v))Q(v)$ will come up as integral values.

Assume that $Q(\tau) =: Q$ is constant along γ, and use the theorem
(4.2.2.). We find $U(\tau_0, \tau_1)(\tilde{\psi}_i(\tau_0)) = \tilde{\psi}_i(\tau_1) = exp[\int_\gamma Q d\tau] \ (\tilde{\psi}_i(\tau_0)) = exp[\int_\gamma \kappa_i(\tau) d\tau] \ (\tilde{\psi}_i(\tau_0))$, and in particular,

$$\frac{\partial \tilde{\psi}_i(\tau)}{\partial \tau} = \kappa_i exp(\int_\gamma \kappa_i(\tau) d\tau)(\tilde{\psi}_i(\tau_0)) = Q(\tilde{\psi}_i(\tau)),$$

so, of course, again the Schrödinger's equation, with τ as time. For an example of a non-constant Hamiltonian, see Example (4.8) and (4.9).

Above, $Simp_n(\mathbf{A}(\sigma))$ is our *time-space*, and $Simp_1(A)$ or $Simp_1(\mathbf{A}(\sigma))$ are the analogues of the classical configuration space. Given an element $v \in Simp_n(\mathbf{A}(\sigma))$, corresponding to a simple module V of dimension n, there are for every $a \in \mathbf{A}(\sigma)$, a set of n possible values, namely its eigenvalues, as operator on V. Since V is simple, the structure map,

$$\rho_V : \mathbf{A}(\sigma) \longrightarrow End_k(V)$$

is supposed surjective, and so in general (and, for order 2 dynamical systems, always) the operators $\tilde{\rho}(a)$ and $\tilde{\rho}(da), a \in A$, cannot all commute. In fact, if $dim_k V = \infty$, or $dim_k V$ is approaching ∞, see Example (4.3) and (4.4), any one $a \in A(\sigma)$ would tend to have a conjugate, i.e. an element $b \in A(\sigma)$, such $[\tilde{\rho}(a), \tilde{\rho}(b)] = \mathbf{1}$. Therefore, if the values q_i of $\tilde{\rho}(a))$ are determined, then the values p_i of $\tilde{\rho}(b)$ will be totally biased, and vice versa, giving us the *Heisenberg uncertainty problem*. In general there is no way of fixing a point of $Simp_1(\mathbf{A}(\sigma))$ as *representing* V or finding natural morphisms,

$$Simp_n(\mathbf{A}(\sigma)) \longrightarrow Simp_m(\mathbf{A}(\sigma)), m < n.$$

However, as we know from Chapter 3, see also [18], there are partially defined *decay* maps,

$$Simp_n(\mathbf{A}(\sigma))^\infty := \overline{Simp_n(\mathbf{A}(\sigma))} - Simp_n(\mathbf{A}(\sigma)) \to \oplus_{n > m \geq 1} Simp_m(\mathbf{A}(\sigma)).$$

In the very special case, where $A = k[x_1, ..., x_p]$ is a commutative polynomial algebra, there exists moreover, for every linear form $\int : V \to k$, and every state $\psi(\tau) \in \tilde{V}|\gamma$ a curve $\Psi(\gamma) \subset Spec(A) \simeq \mathbf{A}^p$ defined, by its coordinates, in the following way,

$$x_{i(\tau)} = \int \tilde{\rho}(x_i) \psi(\tau) / \int \psi(\tau), \ i = 1, .., p.$$

Here $\tilde{V}|\gamma$ is identified with $V \otimes_k O_\gamma$, τ being a parameter of γ. If we are able to pick common eigenfunctions $\{\phi_j \in \tilde{V}_\gamma\}$, $j = 1, ..., n$ for $\tilde{\rho}(x_i)$, $i = 1, ..., p$, generating $\tilde{V}_\gamma$, with eigenvalues $\kappa_j^i(\tau)$, $j = 1, ..., n$, $i = 1, ..., p$, and if

$\psi(\tau) = \sum_j \lambda_j(\tau)\phi_j$, then picking the linear form defined by, $\int \phi_j = 1$, $j = 1, ..., n$, we find,

$$x_i(\tau) = \sum_j \lambda_j(\tau)\kappa_j^i(\tau) / \sum_j \lambda_j(\tau),$$

which is a general form of Ehrenfest's theorem.

Suppose now that we have a situation where there is a unique non-trivial positive (as a real function) Planck's constant, $\hbar \in C(n)$. Consider $f_\hbar \in A(\sigma)$, and assume that there are among the $\{f_\lambda\}_\lambda$ a conjugate, i.e. a f_μ such that $[\tilde{\rho}(f_\hbar), \tilde{\rho}(f\mu)] = \mathbf{1}$. This obviously cannot happen unless $dim_k V = \infty$, but see the Examples (4.3) and (4.4) for what happens at the limit when $dim_k V$ goes to ∞.

Then we easily find that $\mu = -\hbar$. Moreover, if ψ_0 is an eigenvector for Q with least energy (assumed always positive), κ_0, then $N := f_{-\hbar}f_\hbar$ is a *quanta-counting* operator, i.e. $N(\psi_i) = i$, when $\kappa_i = \kappa_0 + (i - 1)\hbar$, is the $i - th$ energy level. It follows also that $[Q, f_{-\hbar}f_\hbar] = 0$. The algebra generated by $\{f_\hbar, f_{-\hbar}\}$ is a kind of a *Fock representation*, $\mathbf{F}$ on a *Fock space*. Its Lie algebra of derivations turns out to contain a *Virasoro*-like Lie-algebra. Since for $Q_h := f_{-\hbar}f_\hbar$ we have that $[Q_h, f_{-\hbar}] = f_{-\hbar}, [Q_\hbar, f_\hbar] = f_\hbar$, it is often seen in physical texts that one identifies the Hamiltonian, Q, with $Q_\hbar$, or with $\sum_\hbar Q_\hbar$. We shall return to this in the Examples (4.4), (4.5) and (4.12), at the end of this Chapter. See also [30] I, (4.2), pp.173-176, and let us pause to explain why Weinberg's (two seemingly different definitions of the) creation and annihilation operators, coincide with our operators, $f_\hbar$, resp. $f_{-\hbar}$, in his case.

This is a consequence of the fact that his momentum operators, p, commute with the Hamiltonian, Q. Therefore the operators $ad(Q)$ and $ad(p)$ commute, and so we may find a common set of eigenvectors for these operators. The result is that the creator operators defined w.r.t. energy, Q, and w.r.t. momentum p, should be physically equivalent.

We have seen that starting with a finitely generated k-algebra A, and a dynamical system σ, we have created an infinite series of spaces $Simp_n(\mathbf{A}(\sigma))$ and a quantum theory, on étale coverings $U(n)$, of these spaces, with time being defined by the Dirac derivation δ.

Each $C(n)$ is commutative and $\tilde{V}$ is a versal bundle on $U(n) \subset Simp_1(C(n))$. Moreover, the elements of $\mathbf{A}(\sigma)$, the observables, become sections of the bundle of operators, $End_{C(n)}(\tilde{V})$.

Clearly, if $D \subset Simp_1(C(n))$ is a subvariety, say a curve parametrized by some parameter q, then the universal family induces a homomorphism of algebras,

$$\tilde{\rho}_D : \mathbf{A}(\sigma) \longrightarrow End_D(\tilde{V}|D).$$

This is in many recent texts referred to as a *quantification* of the commutative algebra $A(\sigma)/[A(\sigma), A(\sigma)]$, or to a *quantification deformation*, and the parameter q is sometimes confounded with Planck's constant. This is unfortunate, but probably unavoidable!

In quantum theory one attempts to treat the *second quantification* of an oscillator in dimension 1, as a certain representation on the Fock space, i.e. constructing observables acting on Fock space, with the properties one wants. This turns out to be related to the canonical representations of $Ph(C) := k < x, dx >$ on an n-bundle over the algebra, $R := k[[n]_{p,q}]$. Here the p, q-*deformed numbers* $[n]_{p,q}$ are introduced as,

$$[n]_{p,q} := q^{n-1} + pq^{n-2} + p^2 q^{n-3} + ... + p^{n-2}q + p^{n-1},$$

and we may as well consider p, q as formal variables, so that $R \subset k[p, q]$.

See Example (4.5) where we construct a homomorphism of $A(\sigma)$ into an endomorphism ring of the form $End_R(V \otimes_k R)$, see ([2], Appendice, on the *q-commutators*). Picking representatives for x and dx in $M_n(R)$, it turns out that, instead of the classical defining relations for an oscillator, i.e. with $a_+ := x + dx$, $a_- : x - dx$, and with a Hamiltonian Q, such that in $End_R(V \otimes_k R)$,

$$[Q, x] = dx, \ [Q, dx] = x, \ [a_-, a_+] = 1$$

one obtains,

$$[Q, x]_q = dx, \ [Q, dx]_q = x, \ [a_-, a_+]_q = 1$$

where $[a, b]_q := ab - qba$ is the *quantized* commutator. This holds in particular for $p = 1$, so for $R = k[q]$, defining a curve D in $Simp_n(Ph(k[x]))$.

However, this $k[q]$-parametrization is not parametrizing an integral curve of ξ in $Simp_n(Ph(C))$. On the contrary, it is parametrizing a curve of *anyons* with $q = -1, 1$ corresponding to, respectively, Fermions and Bosons. A simple computation shows that it is transversal to ξ, and therefore represent a phenomenon which takes place instantaneously, see the Examples (4.4), (4.5).

4.5 General Quantum Fields, Lagrangians and Actions

Given algebras A and B, supposed to be moduli spaces of interesting objects. Given dynamical structures (say of order 2) (σ) and (μ) of $Ph^{\infty}(A)$

and $Ph^\infty(B)$ respectively, corresponding to Dirac derivations, and corresponding vector fields, $\delta_\sigma, \xi_\sigma$ and δ_μ, ξ_μ, respectively, we consider the set (or space, see Chapter 3), $\mathbf{F}(A, B)$, of iso-classes, of morphisms $\phi : A \to B$. Any such induces a morphism,

$$Ph(\phi) : PhA \to PhB.$$

and we shall assume also a morphism,

$$\phi : A(\sigma) \to B(\mu).$$

This is actually what we shall call a *field*. The space of such is denoted $\mathbf{F}(A, B)$.

The meaning of the term *field*, or its physical interpretation, is not obvious. In standard physics texts the notion is rarely well defined, see e.g. [30], I, (1.2), where one finds a nice historical introduction, and good reasons for lots of mathematical tears!

I would like to consider $\phi : A \to B$ as a morphism of the moduli space $Spec(B)$, of *physical objects* Y, into the moduli space $Spec(A)$, of *physical objects* X, and in this way modeling composite physical objects (X, Y), as we shall see below.

Now, apply the deformation theory of categories of algebras, see e.g. [14]. From this theory follows readily that the tangent space of $\mathbf{F}(A, B)$ at a point, ϕ, is,

$$T_{\mathbf{F}(A,B),\phi} = Der_k(A(\sigma), B(\mu))/Triv,$$

where, as usual, $Triv$ depends upon the definition of $\mathbf{F}(A, B)$, i.e. upon the notion of isomorphisms among fields, see [14].

Unfortunately, this moduli space, $\mathbf{F}(A, B)$, is not, in general, a prescheme, neither commutative nor non-commutative. As we have, as a rule in this paper, identified any k-algebra with some *space*, we shall, never the less, at this point not hesitate to identify $\mathbf{F}(A, B)$ with the k-algebra of (reasonable) functions (or operators) defined on the space, and denote it by, $F(A, B)$. Then we are free to consider the versal (or maybe, universal) *family of quantum fields*,

$$\tilde{\phi} : A \to F(A, B) \otimes_k B.$$

Just in the same way as above, there is now a canonical vector field $[\delta]$ on the space $\mathbf{F}(A, B)$, defined by its value at ϕ, given by,

$$[\delta](\phi) = cl(\delta_\sigma \phi - \phi \delta_\mu).$$

The set of stable, or singular fields, is now in complete analogy with the singular points in $Simp(A(\sigma))$ mentioned above, and treated in detail in the examples, (3.5)-(3.9) below,

$$\mathbf{M}(A, B) := \{\phi \in \mathbf{F}(A, B)|[\delta](\phi) = 0\}.$$

Here one sees where the *Noether function* Q enters. In fact, if we are identifying fields up to automorphisms of B defined by trivial derivations, $[\delta](\phi) = 0$ is equivalent to the existence of a Hamiltonian $Q \in B(\mu)$, such that for every $a \in A(\sigma)$

$$\delta_B(\phi(a)) - \phi(\delta_A(a)) = Q\phi(a) - \phi(a)Q = [Q, \phi(a)].$$

Consider this equation in rank 1, i.e. look at the commutativizations

$$Ham : A(\sigma) \to A(\sigma)/[A(\sigma), A(\sigma)] =: A_0(\sigma).$$
$$Ham : B(\mu) \to B(\mu)/[B(\mu), B(\mu)] =: B_0(\mu),$$

We find that in $Simp_1(B(\mu))$ the equation above reduces to,

$$\delta_B(\phi(a)) = \phi(\delta_A(a)),$$

which, geometrically, means the following: If γ is an integral curve of δ_A, in $U^A(n)$, then the inverse image via ϕ is a union of integral curves for δ_B in $U^B(n)$.

The actual definition of a dynamical structure (σ) has, up to now, been loosely treated. It may be defined in terms of *force laws* i.e. where (σ) is the two-sided δ-stable ideal generated by the equations $(\delta^n t_p - \Gamma^p)$, where,

$$\delta^n t_p := d^n t_p, \ \Gamma^p(\underline{t_i}, \underline{dt_j}, \underline{d^2 t_k}, .., \underline{d}^{n-1} t_l) \in Ph^\infty(A), \ p = 1, 2, ..., d.$$

But, in general, force laws like these don't pop up naturally. In fact, Nature seems to reveal its structure through some Action Principles. The physicists are looking for a Lagrangian L, and an action functional $S(L)$ defined on $\mathbf{F}(A, B)$. In our setting, L is simply an element,

$$L \in Ph(A).$$

For every field $\phi \in \mathbf{F}(A, B)$, the *action*, usually denoted,

$$S(\phi) := S(L)(\phi) \in k,$$

is constructed via some particularly chosen representation,

$$\rho : B(\mu) \to End_k(V).$$

Put, $\mathbf{L} := \rho(\phi(L))$ and let,

$$S(\phi) := Tr(\mathbf{L}).$$

In the classical case the trace, Tr, is an integral.

We may choose several canonical representations ρ, like the versal family of rank 1 representations, treated above, and called,

$$Ham : Ph(B) \to Ph_0(B) := Ph(B)/[Ph(B), Ph(B)],$$

or the corresponding in rank n. We may also, for any derivation ζ of B, consider the canonical homomorphism $\rho_\zeta : Ph(B) \to B$, as a representation. In the first case it is clear that $Tr(\mathbf{L})$ is simply the image of $\mathbf{L}$ in $Ph_0(B)$. In the last case the *Lagrangian density*, i.e. $\mathbf{L}$, is now a function, $\mathbf{L}(\phi_i, \zeta(\phi_j))$, in $\phi_i := \phi(t_i)$, and in $\zeta(\phi_j)$ for some generators t_i of A, and Tr is an integral over some reasonably well defined subspace of $Simp_1(B)$. In this case one usually has to impose some boundary conditions on ϕ.

Clearly, $S := S_\rho = Tr(\mathbf{L})$ is a function,

$$S : \mathbf{F}(A, B) \to k,$$

and $\nabla S \in \Theta_{\mathbf{F}(A,B)}$, is a vector field that corresponds to the fundamental vector field $\xi = [\delta]$, above. The equations defining the singularities of ∇S, is usually written,

$$\delta S := \delta \int \mathbf{L} = 0,$$

since for most classical representations the dimension of V is infinite, and the trace is an integral, see examples below.

Here is where the calculus of variation comes in. The corresponding Euler-Lagrange equations, the *equations of motion*, picks out the set of solutions, *the singular fields*, i.e. $\mathbf{M}(A, B) \subset \mathbf{F}(A, B)$.

The subspaces $\mathbf{M}(A, B)$ in $\mathbf{F}(A, B)$, defined by the Euler-Lagrange equations, are therefore uniquely defined by L, without reference to any dynamical structures of A or B..

The problem with this is that, unless there actually exist a dynamical structure corresponding to ∇S, we cannot know that our *laws of nature* are mathematically deterministic, see the Introduction, and compare [30], I, chapter 7.

Notice that the classical field theory corresponds to the situation where $A = k[\underline{t}]$ and $B = k[\underline{x}]$, and where ϕ is defined in terms of the *fields*, $\phi_i := \phi(t_i)$, and their *time derivatives* $\dot{\phi}_i := \phi(dt_i) := d\phi_i \in Ph(B)$. Choosing the representation ρ_ζ for some derivation ζ, of B, we may assume the *Lagrangian density* has the form,

$$\mathbf{L} := L(\phi_i, \phi_{j,\alpha})$$

where $\phi_{j,\alpha} := \frac{\partial \phi_j}{\partial x_\alpha}$. The singular fields, are then picked out by the variation of the action integral, i.e.

$$\delta \int \mathbf{L} d\mu = 0,$$

or by the corresponding Euler-Lagrange equations,

$$\frac{\partial \mathbf{L}}{\partial \phi_i} - \sum_\alpha \frac{\partial}{\partial x_\alpha}(\frac{\partial \mathbf{L}}{\partial \phi_{i,\alpha}}) = 0.$$

In the light of the above, considering these equations as *equations of motion*, is now, maybe, a reasonable guess.

In fact, we shall show that this *Lagrangian* theory actually produce a *dynamical structure*, at least in special cases.

Pick a Lagrangian $L \in Ph(A)$, and assume $B = M_n(k)$, so that $\mathbf{F}(A,B) = Rep_n(A) \subset Rep_n(Ph(A))$. Restrict to $Simp_n(Ph(A)) \subset Rep_n(Ph(A))$, and consider the versal family,

$$\tilde{\rho} : Ph(A) \to M_n(C(n)).$$

Put $\mathbf{L} := \tilde{\rho}(L) \in M_n(C(n))$ and $S := Tr\mathbf{L} \in C(n)$. If the choice of the Lagrangian L, is clever, the gradient, $\nabla S \in \Theta_{C(n)}$, restricted to $U(n)$ is a candidate for the vector field $\xi = [\delta]$, induced by the Dirac derivation δ defined by some dynamical structure, $\mathbf{A}(\sigma)$. If the philosophy of contemporary physics is consistent, this is what we would expect.

Based on the parsimony principle involved in the theory of Lagrange, and given a dynamical system, with Dirac derivation δ, we should expect that the Lagrangian L is constant in time, i.e. that we have the *Lagrangian equation*,

$$\delta(L) = 0.$$

But then Theorem (3.4) tells us that, in the situation above, we have in $C(n)$,

$$[\delta](Tr\mathbf{L}) = 0,$$

i.e. for all $n \geq 1$, the equation $\nabla S = 0$, picks out the solutions γ of the theory. Now we may try to turn the argument upside down, and ask whether, given L, we may construct a Dirac derivation, δ, from the *Lagrangian equation* above. This is the purpose of the next examples. But be aware, this is not proving that the Lagrangian method for studying quantum fields, is equivalent to the one I propose above. Example (4.4) shows that there exists simple Lagrangians L inducing unique force laws, but such that the set of solutions $\{\gamma\}$ is not determined by $ker(\delta)$. Moreover, the classical relation between the Lagrangian and the Hamiltonian turns out to be more subtle in this non-commutative case. Notice that, above, we have accepted *fields* $\phi : A \to B$ where $B = M_n(k)$ is a finite non-commutative *space*, for which the usual Euler-Lagrange equations do not apply.

Solving differential equations like the Lagrange equation, in non-commutative algebras, is not easy. However, if we reduce to the corresponding commutative quotient, things become much easier. In fact, as we mentioned in the Introduction, in the commutative situation we may write, in $Ph(A)$,

$$\delta = \sum_i (dt_i \frac{\partial}{\partial t_i} + d^2 t_i \frac{\partial}{\partial dt_i}),$$

and the Lagrange equation will produce order 2 dynamical structures, see Example (4.1). We may also consider the Euler-Lagrange equations, impose δ, as time, and solve,

$$\delta(\frac{\partial L}{\partial dt_i}) - \frac{\partial L}{\partial t_i} = 0,$$

to find an order 2 force law, $d^2 t_i = \Gamma(t_i, dt_j)$.

The strategy will be to solve the equation in a representation like Ham, then try to lift it to $Ph(A)$ and then, eventually, map it back to say $Wey : Ph(A) \to Diff_k(A, A)$. We shall now show that this strategy works in some interesting cases.

But let us first have a look at the relationship, as we see it, between the picture we have drawn of QFT, and the one physicists presents in modern university textbooks.

4.6 Grand Picture: Bosons, Fermions, and Supersymmetry

Consider a situation with a dynamical system, with Dirac derivation δ, and fix the rank n versal family,

$$\mathbf{A}(\sigma) \to End_{C(n)}(\tilde{V}).$$

Look at the singularities of the fundamental vectorfield $\xi \in Der_k(C(n))$. Let V be a corresponding representation, the *particle*. Compute the set of eigenvalues Λ of adQ acting on $End_k(V)$, and the set of minimal elements, i.e. the set of Planck's constants, $\{\hbar\}$, and the corresponding eigenvectors $f_\hbar \in End_k(V)$. We shall see in Examples (4.3) and (4.4), that if there exists a conjugate to $f_\hbar$, it must be $f_{-\hbar}$. If the Hamiltonian, Q, is diagonalized, with eigenvalues $\{q_0 \le q_1 \le \dots \le q_{n-1}\}$ it is of course easy to see that $\Lambda = \{(q_i - q_j)\}_{i,j=0,\dots,n-1}$, $f_\lambda = \sum_{q_i - q_j = \lambda} \epsilon_{i,j}$, and in particular, $f_{-\hbar} = f_\hbar^*$, the conjugate matrix.

Anyway, choosing a *vacuum state* $\phi_0 \in V$ for the Hamiltonian Q, i.e. an eigenvector with minimal positive-or zero-eigenvalue, we find that, for some $i \ge 0$, unless $f_\hbar^i = 0$, we have $Q(f_\hbar^i(\phi_0)) = i\hbar f_\hbar(\psi_0)$, i.e. the state $\phi_i := f_\hbar^i(\phi_0)$ may be *occupied* by i quantas. If $\{\phi_i\}_{i=0,\dots,n-1}$ is a basis for V then this is the purely *Bosonic case*, with $q_i - q_{i-1} = \hbar$, see (4.4), where we have treated the simple case of the harmonic oscillator.

What do I mean by *state occupied by several quantas*? The language is far from clear. Here we shall restrict ourself to the elementary language of quantum physics,. The phrase, *the state ϕ is occupied by n quanta* shall mean that ϕ is an eigen-state of the Hamiltonian Q, with eigen-value $n\hbar$. We shall also, as is explained above, assume $\hbar = 1$.

Physicists have come to the realization that there exist two types of *particles*, Bosons and Fermions, with different *statistics*, in the sense that states *containing* several identical Bosons are invariant upon permutations of these, but states *containing* several identical Fermions change sign with the permutation. This is another way of expressing that Fermions, like electrons, cannot all sink in to the lowest energy state in an atom, and stay there, killing chemistry.

We shall delay the discussion of collections of identical particles to Chapter 5.

Bosons can have states with an arbitrary occupation number, but Fermions have states only with occupation numbers 0 or 1.

If we know that no states are *occupied* by more than one quantum *at a time*, then we must conclude that,

$$f_\hbar^2 = f_{-\hbar}^2 = 0.$$

Moreover, we pose,

$$\{f_\hbar, f_{-\hbar}\} := f_{-\hbar} f_\hbar + f_\hbar f_{-\hbar} = 1,$$

implying, $(f_\hbar + f_{-\hbar})^2 = 1$. This is the purely *Fermionic case*.

These relations induce a split-up of the representation V, i.e.

$$V \simeq V_0 \oplus V_1,$$

In fact, put $R_0 := f_\hbar f_{-\hbar}$, $R_1 := f_{-\hbar} f_\hbar$, and see that

$$R_0 + R_1 = 1, \ R_0 R_1 = R_1 R_0 = 0, R_i^2 = R_i, i = 0, 1,$$

and put $V_0 = imR_0, V_1 = imR_1$. Since R_i is the identity on V_i, it is clear that the two linear maps,

$$f_{-\hbar} : V_0 \to V_1, \quad f_\hbar : V_1 \to V_0,$$

are isomorphisms, thus $dim_k V_0 = dim_k V_1 = 1/2\, n$. Clearly, any endomorphism of V can be cut up into a sum of graded endomorphisms. Those of degree 0 we would like to call Bosonic. Those of degree 1, or -1, should then be called Fermionic. In dimension 4, this would look like:

$$Q = \begin{pmatrix} q_{1,1} & q_{1,2} & 0 & 0 \\ q_{2,1} & q_{2,2} & 0 & 0 \\ 0 & 0 & q_{1,1}+1 & q_{1,2} \\ 0 & 0 & q_{2,1} & q_{2,2}+1 \end{pmatrix},$$

with,

$$f_{-\hbar} = \begin{pmatrix} 0 & 0 & 1 & 0 \\ 0 & 0 & 0 & 1 \\ 0 & 0 & 0 & 0 \\ 0 & 0 & 0 & 0 \end{pmatrix} \text{ and } f_\hbar = \begin{pmatrix} 0 & 0 & 0 & 0 \\ 0 & 0 & 0 & 0 \\ 1 & 0 & 0 & 0 \\ 0 & 1 & 0 & 0 \end{pmatrix}$$

In the general case we may have a mix of Bosons and Fermions present, and this leads to the notion of *super symmetry*.

If we have a split up, as above, the modules $V_i, i = 1, 2$, having the Hamiltonians Q_0, and $Q_1 := Q + 1$, implying that, as Hamiltonians, $Q_o = Q_1$, we see that $End_k(V)$ is generated by the Bosonic operators, $a := f_\hbar$, $a^+ := f_{-\hbar} \in End_k(V_i)$, both defined for $Q_0 = Q_1$, together with the Fermionic operators $f^- := f_\hbar \in Hom_k(V_1, V_0)$, $f^+_{-\hbar} \in Hom_k(V_0, V_1)$, for the Hamiltonian Q. In fact,

$$End_k(V) = \begin{pmatrix} End_k(V_0) & Hom_k(V_0, V_1) \\ Hom_k(V_1, V_0) & End_k(V_1) \end{pmatrix}.$$

is generated by the $End_k(V_0)$ and $End_k(V_1)$, together with the isomorphisms $f_{-\hbar} := (V_0 \to V_1) \in Hom_k(V_0, V_1)$, $f_\hbar := (V_1 \to V_0) \in Hom_k(V_1, V_0)$.

Put, $f = \imath(f^- + f^+)$, and see that there are two eigenstates of f, the Fermion with eigenvalue 1, and the *anti-Fermion* with eigenvalue -1.

The general situation is much like the one above. We may assume that we have a Hamiltonian Q, split up as above, with corresponding Bosonic operators, a_l, a^+_l, and Fermionic operators, f_p, f^+_p, generating $End_k(V)$. We may also assume that, given the vacuum state $\phi_0 \in V$, there is a basis of V, given by $\{\phi_i := (a^+_l)^i(\phi_0)\}_{i=0}^{n-1}$. Moreover a_l kills ϕ_0, and a^+_l

kills ϕ_{n-1}. Considering the versal family $\tilde{\rho}$, and the global Hamiltonian $Q \in End_{C(n)}(\tilde{V})$, the situation becomes more subtle. Here we have the étale morphism $\pi : U(n) \rightarrow Simp_n(A(\sigma))$. Fix the field $k = \mathbf{R}$, and assume no harm is made by this choice. Clearly we have a monodromy homomorphism,

$$\mu(v) : \pi_1(v; Simp_n(A(\sigma))) \rightarrow Aut(V) \simeq Gl_n(\mathbf{R}).$$

One would be tempted to define Bosons, Fermions, and Anyons, with respect to this monodromy map. The component of $Simp_n(A(\sigma))$ where $\mu(v)$ is trivial are Bosonic, the one with $im(\mu(v)) = \{+1, -1\} \simeq \mathbf{Z}_2$ is Fermionic, and the rest are Anyonic. Notice that the fiber of π is composed of *identical particles*. The treatment of such are, as mentioned above, postponed until Chapter 5.

If Q is constant, we may of course assume that the Bosonic operators, a_l, a_l^+, and Fermionic operators, f_p, f_p^+, are elements of $End_{C(n)}(\tilde{V})$, generating $End_{C(n)}(\tilde{V})$, as $C(n)$-module. Then any quantum field would look like,

$$\psi(v) = \psi(v, \underline{a}, \underline{a}^+, \underline{f}, \underline{f}^+),$$

the functions being polynomials in the operator variables. In particular $\tilde{\rho}(t_j)$ and $\tilde{\rho}(dt_j)$ would have this form, where, in most cases relevant for physics, the polynomial function would be linear, see the case of the harmonic oscillator, Example (4.4).

This is very close to the form one finds in physics books, the only problem is that the function ψ is a function on $Simp(\mathbf{A}(\sigma))$, not on the configuration space, with fixed momentum, as is usually the case in physics.

Suppose we have a classical case, where the algebra $A = k[t_1, ..., t_r]$ is the commutative affine algebra of the *configuration* variety, $X := Spec(A)$. Then an element $v \in Simp_n(\mathbf{A}(\sigma))$ will correspond to up to n different points in $q_i \in X$. If one imposes commutation rules on the dt_j, as physicists do, then to v, there corresponds also up to n values of the momenta, $p_l \in Spec(k[dt_1, ..., dt_r])$. However, there is no way to pinpoint the representation v, by fixing q and p. Because t_i and dt_i do not commute, which imposes the Heisenberg uncertainty relation with respect to determining $q's$ and $p's$, at the same time, the physicists will have to introduce some mean values, using different versions of the spectral theorem for Hermitian operators, to obtain reasonable definitions of the notion of quantum field. Usually the generators, of the algebra of quantum fields, are expressed in the form of an integral, like,

$$\psi(x) = \sum_{\sigma} (2\pi)^{-3/2} \int d^3p [u(p,\sigma,n)a(p,\sigma,n)exp(ipx) \qquad (4.1)$$

$$+ v(p,\sigma,n^c)a^+(p,\sigma,n)exp(-ipx)], \qquad (4.2)$$

see [30], I, p.260. Here σ means spin, n the particle *species*, and n^c the antiparticle of the species n. Integration is on different domains, depending on the situation. and the whole thing is deduced from ordinary quantum theory, imposing relativistic invariance.

The corresponding interpretation of interaction, implies that interaction takes place at *points* in configuration space. This is the so called *locality of action*. See the very readable article of Gilles Cohen-Tannoudji in [11], p.104.

Of course, the interesting Hamiltonians $Q \in End_{C(n)}(\tilde{V})$ will not be constant, therefore physicists introduce what is called *perturbation theory*, which amounts to assuming that there exists a background situation, defined by an essentially constant Hamiltonian Q_0, such that for the real situation, given by the versal family $\tilde{\rho} : \mathbf{A}(\sigma) \rightarrow End_{C(n)}(\tilde{V})$, and the Dirac derivation δ, the Hamiltonian Q may be considered as a perturbation of Q_0, with an *interaction* $I \in End_{C(n)}(\tilde{V})$, such that,

$$Q = Q_0 + I.$$

Then, using the basis $\{\phi_i\}$, given for $\tilde{V}$, defined by the *creation operators* $\{a_l^+\}$, see above, one may apply Theorem (4.2.2), and obtain formulas for the evaluation operator, along the curve γ, applied to any ϕ_i. If we have a Hermitian metric on the bundle $\tilde{V}$, then we obtain formulas for the so called $S = (S_{i,j})$-matrix, calculating the probability for a ϕ_i observed at the start-point v_0 of γ, to be observed as changed into ϕ_j, at the end-point v_1. The same types of formulas as one finds in elementary physics books, like [30], I, p.260, pop up. And the computation is again made easier by chopping up the formulas, by introdusing Feynman diagrams. In our case, the integrals along the compact γ, are, of course, easily seen to be well defined, but then we have not explained how we may know that our preparation gave us the start-point v_0.

This is where the problem of locality of action enters. Suppose we have fixed a basis, $\{\phi_i\}_{i=0}^{n-1}$, of the $C(n)$-module of sections of $\tilde{V}$, composed of common eigenvectors for the commuting operators, $\tilde{\rho}(t_j) \in End_{C(n)}(\tilde{V})$. Suppose also that the operator $[\tilde{\rho}(t_j), \tilde{\rho}(dt_j)]$ is *sufficiently close to* the identity, or rather, strictly bigger than zero on a compact part of $Simp_n(\mathbf{A}(\sigma))$, see Examples (4.4) and (4.5). Notice again that it must have a vanishing trace, since we are in finite dimension. Fix an index l and let $\{\kappa_i(l)\}_{i=0}^{n-1}$ be a basis of the $C(n)$-module of sections of $\tilde{V}$, composed of eigenvectors for the operator $\tilde{\rho}(dt_l)$. Then, given a $v \in Simp_n(\mathbf{A}(\sigma))$, represented by the n-dimensional $\mathbf{A}(\sigma)$-module V, we have,

$$t_j(\phi_i) = q_i^j \phi_i, \ dt_l(\kappa_j(l)) = p_j^l \kappa_j(l),$$

where, for $i = 0, ..., n-1$, $\underline{q}_i(v) := \underline{q}_i = (q_i^1, ..., q_i^r) \in X$ are the possible configuration positions of v, and, for $j = 0, ..., n-1$, the possible values of the l-component of the momenta, are given by, $p_j^l(v) := p_j^l$. Consider now the base-change matrices, (λ_i^j), and (μ_i^j), such that, $\phi_j = \sum \lambda_i^j \kappa_i$, and $\kappa_j = \sum \mu_i^j \phi_i$, and compute $[dt_l, t_l](\phi_i)$. We obtain,

$$[dt_l, t_l](\phi_i) = \sum_{j,k} \lambda_j^i p_j^l \mu_k^j (q_i^l - q_k^l) \phi_k.$$

By assumption, the base change matrices, (λ_i^j), and (μ_i^j), must be bounded on the compact subset of $Simp_n(\mathbf{A}(\sigma))$, and the operator $[dt_l, t_l]$ is not trivial in the same subset. This implies that when the l-coordinate of the configuration positions are clustered tightly about a certain point, then the l-coordinate of the corresponding momenta cannot be kept bounded. This is the analogy of the *Heisenberg uncertainty relation* of the classical quantum theory.

With this in mind, one would be tempted to formulate the task of the experimenter in physics, as follows.

She should test out the possibilities of the laboratory technology, to prepare the situation by bounding the configuration positions of the phenomenon she is interested in, to a subset $D(q) \subset X := Spec(k[t_1, ..., t_r])$, and at the same time bound the corresponding l-component of the momenta, to a subset $D(p, l) \subset Y := Spec(k[dt_1, ..., dt_r])$ by performing repetitive experiments. Each experiment, setting up the preparation would have to be performed within a short time interval $\Delta\tau$. Then she should compute the subset,

$$D(q, p, l) = \{v \in Simp(\mathbf{A}(\sigma)) | \underline{q}_i(v) \in D(q), \ p_j^l(v) \in D(p, l), \ i, j = 1, ..., r\},$$

and finally, she should compute for each $v \in D(q, p, l)$ the solution curve $\gamma_v(\tau)$ through v with length τ, that is, with time-development τ, ending at $v(\tau)$. The end-points of all of these curves, would form a subset $D(q, p, l : \tau)$, and one would expect that the result of letting the phenomenon develop through the time-interval τ, would give position and momenta results within the subsets,

$$D(q:\tau) = \{\underline{q}_i(v) | v \in D(q, p:\tau)\}, \quad D(p, l:\tau) = \{\underline{p}_i^l(v) | v \in D(q, p, l:\tau)\}.$$

The philosophically interesting result would be that no interaction is really local.

One interesting consequence of the above assumption, our Heisenberg uncertainty relation, is that if we are considering a natural phenomenon related to a macroscopic object, i.e. such that all $|q_k^l - q_i^l|$ are allowed to be, relatively, very big, then we may prepare the object in such a way that all $|p_k^l - p_i^l|$ are very small. We then have a classical situation, where the result would be, relatively, unique! The Big Bang, see the last subsection of this Chapter, would in this respect, be the extreme opposite situation, where we are totally incapable to trace unique curves, γ, from the assumed unique point in *configuration* space where BB happens. And, of course, the End of it all, would correspond to a totally homogenous universe, with a uniquely given future!

Example 4.1. Let C be a finite type commutative k-algebra, say parametrizing an interesting moduli space, and assume it is non-singular, and pick a system of regular coordinates $\{t_1, t_2, ..., t_r\}$ in C. The problem of constructing a dynamical system of interest to physics, has been discussed in the Introduction, and above. We may consider an element $L \in Ph(C)$, a Lagrangian, and try to find a force law, with Dirac derivation δ, such that,

$$\delta(L) = 0.$$

We could start with the trivial Lagrangian, $L := g = \sum_{i=1,..,r} dt_i^2 \in PhC$. The Lagrange equation becomes, $0 = \delta(g) = \sum_{i=1,..,r} (d^2 t_i dt_i + dt_i d^2 t_i) \in PhC$. with the obvious solution,

$$d^2 t_i = 0, \quad i = 1, ..., r.$$

inducing a dynamical structure (σ) in $Ph(C)$, generated by the relations,

$$[dt_i, t_j] + [t_i, dt_j], [dt_i, dt_j], i \neq j.$$

The corresponding dynamic system, $C(\sigma)$, is the dynamical system for a *free particle*. Notice however, that, classically, one imposes also the relations, $[dt_i, t_j] = 0$ for $i \neq j$, and $[dt_i, t_i] = 1$.

Consider the representations of dimension 1, corresponding to $\rho = Ham$, and use Theorem (3.2), with n=1. Then, obviously, the Hamiltonian Q must be a function, and we find,

$$\tilde{\rho}(dt_i) = [\delta](t_i), 0 = \tilde{\rho}(\delta^2(t_i)) = [\delta]([\delta](t_i)).$$

This fits well with,

$$[\delta] = \sum_{i=1}^{r} dt_i \frac{\partial}{\partial t_i},$$

which gives us the canonical symplectic structure on the commutativization of $Ph(C)$, the $C(1)$ for this situation. Notice that the corresponding Poisson bracket now give us,

$$\{dt_i, t_j\} = \delta_{i,j},$$

defining a *deformation* of the commutative phase space which is the quotient of $C(\sigma)$ defined above.

4.7 Connections and the Generic Dynamical Structure

Now, let, $L := g = 1/2 \sum_{i,j=1,..,r} g_{i,j} dt_i dt_j \in PhC$, be a Riemannian metric. Recall the formula for the Levi-Civita connection,

$$\sum_l g_{l,k} \Gamma^l_{j,i} = 1/2 \left(\frac{\partial g_{k,i}}{\partial t_j} + \frac{\partial g_{j,k}}{\partial t_i} - \frac{\partial g_{i,j}}{\partial t_k} \right).$$

Since,

$$\delta(g) = \sum_{i,j,k=1,..,r} \frac{\partial g_{i,j}}{\partial t_k} dt_k dt_i dt_j + \sum_{i,j,=1,..,r} g_{i,j} (d^2 t_i dt_j + dt_i d^2 t_j),$$

we may plug in the formula,

$$\delta^2 t_l = -\Gamma^l := -\sum \Gamma^l_{i,j} dt_i dt_j.$$

on the right hand side, and see that we have got a solution of the Lagrange equation,

$$\delta(L) = 0,$$

in the commutative situation. This solution has the form of a *force law*,

$$d^2 t_l = -\Gamma^l := -\sum \Gamma^l_{i,j} dt_i dt_j,$$

generating a dynamical structure $(\sigma) := (\sigma(g))$ of order 2. The dynamic system is, of course, as an algebra,

$$\mathbf{C}(\sigma) = k[\underline{t}, \underline{\xi}]$$

where ξ_j is the class of dt_j. The Dirac derivation now has the form,

$$\delta = \sum_l (\xi_l \frac{\partial}{\partial t_l} - \Gamma^l \frac{\partial}{\partial \xi_l}),$$

and the fundamental vector field $[\delta]$ in $Simp_1(\mathbf{C}(\sigma)) = Spec(k[t_i, \xi_j])$, is, of course, the same. Use Theorem (4.2.3),(ii), and see that $[\delta](g) = 0$, which means that g is constant along the integral curves of $[\delta]$ in $Simp_1(Ph(C))$, which projects onto $Simp_1(C)$ to give the geodesics of the metric g, with equations,

$$\ddot{t}_l = -\sum_{i,j} \Gamma^l_{i,j} \dot{t}_i \dot{t}_j.$$

Put $\delta_i := \frac{\partial}{\partial t_i}$, and consider the Levi-Civita-connection,

$$\nabla : \Theta_C \longrightarrow End_k(\theta_C)$$

expressed in coordinates as,

$$\nabla_{\delta_i}(\delta_j) = \sum_l \Gamma^l_{j,i} \delta_l$$

Classsically we define the curvature tensor $R_{i,j}(\delta_k) = \sum_l R^l_{i,j,k}\delta_l$, of a connection ∇, as the obstruction for ∇ to be a Lie-algebra homomorphism. We find,

$$([\nabla_{\delta_i}, \nabla_{\delta_j}] - \nabla_{[\delta_i,\delta_j]})(\delta_k) = \sum_l R^l_{i,j,k}\delta_l.$$

This, we shall see, is a commutative version of the more precise notion of *curvature*, related to a more general dynamic system, to be studied below. Recall that the Ricci tensor is given as,

$$Ric_{i,k}(g) = \sum_j R^j_{i,j,k}$$

and that, assuming the metric is non-degenerate with inverse $g^{k,i}$, one defines the scalar curvature of g, as,

$$S(g) := \sum_{k,i} g^{k,i} Ric_{i,k}.$$

These are fundamental metric invariants. Recall also Einstein's equation,

$$Ric - 1/2S(g)g = U,$$

where U is the *stress-mass* tensor.

A non-degenerate metric, $g \in Ph(C)$ induces an isomorphism of C-modules

$$\Theta_C = Hom_C(\Omega_C, C) \simeq \Omega_C.$$

Assume first that $g = 1/2 \sum_{i=1}^{d} dt_i^2$, i.e. assume that the space is Euclidean, and pick a basis $\{\delta_i := \frac{\partial}{\partial t_i}\}$ of Θ_C, and a basis $\{dt_j\}$ of Ω_C such that,

$$\delta_j(dt_i) = \delta_{i,j}.$$

Consider a C-module, V. Any connection ∇ on V induces a homomorphism,

$$\rho := \rho_\nabla : Ph(C) \rightarrow End_k(V),$$

with, $\rho(dt_i) := \nabla_{\delta_i} = \frac{\partial}{\partial t_i} + \nabla_i$. To see this we just have to check that the relations, $[dt_i, t_j] + [t_i, dt_j] = 0$, in $Ph(C)$ are not violated. Since we obviously have,

$$\rho([dt_i, t_j]) = [\nabla_{\delta_i}, \rho(t_j)] = \delta_{i,j},$$

the homomorphism ρ is well defined. We are therefore led to consider the dynamical structure on C, generated by the ideal,

$$(\sigma) := ([dt_i, t_j] - \delta_{i,j}) \subset Ph^\infty(C).$$

Since $\delta(t_i) = [g, t_i] = dt_i$, the Dirac derivation is given by,

$$\delta = ad(g).$$

(σ) is clearly invariant under isometries. Moreover, in $C(\sigma)$ we have,

$$\delta^2(t_i) = -1/2 \sum_k (dt_k[dt_i, dt_k] + [dt_i, dt_k]dt_k).$$

Notice that if $d^2 t_i = 0$ for i=1,..,d, then $[dt_i, dt_j] = 0$, for all i,j. This will also be true for any constant metric.

Given any connection, ∇, on an C-module, V, and consider the corresponding representation, $\rho : Ph(C) \rightarrow End_k(V)$. If V is of infinite dimension as k-vector space, we cannot prove that there is a useful moduli space in which V is a point. However we now know that ρ is singular. This follows since there exist a Hamiltonian, $Q := \rho(g) \in End_k(V)$, such that for all $a \in Ph(C)$,

$$\rho(da) = [Q, \rho(a)].$$

In particular we have, $\rho(dt_i) = \nabla_{\delta_i} = [Q, t_i]$. This imply,

$$Q = 1/2 \sum_i \nabla_{\delta_i}^2.$$

Thus for any connection ∇ we find a *force law*, in $End_k(V)$, given by,

$$\rho_\nabla(d^2 t_i) = -1/2 \sum_{j=1}^{d} \nabla_{\delta_j}[\nabla_{\delta_i}, \nabla_{\delta_j}] - 1/2 \sum_{j=1}^{d} [\nabla_{\delta_i}, \nabla_{\delta_j}]\nabla_{\delta_j}.$$

We shall in this situation use the notations,

$$R_{i,j} := [dt_i, dt_j] \in Ph(C), \quad F_{i,j} := [\nabla_{\delta_i}, \nabla_{\delta_j}] \in End_C(V),$$

$F_{i,j}$ being the curvature *tensor*, of the connection. Below we shall come back to these notions in the general situation.

Since we now have,

$$\nabla_{\delta_j} F_{i,j} = F_{i,j} \nabla_{\delta_j} + \left(\frac{\partial F_{i,j}}{\partial t_j} + [\nabla_j, F_{i,j}] \right)$$

we find the following equation,

$$\rho_\nabla(d^2\underline{t}) = -F\rho(\underline{dt}) - \underline{q},$$

where $\underline{q}$ (by definition) is the *charge* of the field. $\underline{q}$, is a vector, the coordinates of which,

$$q_i = 1/2 \sum_{j=1}^{r} \left(\frac{\partial F_{i,j}}{\partial t_j} + [\nabla_j, F_{i,j}] \right),$$

are endomorphisms of the bundle. See Example (3.16).

Suppose now that we have a free field, i.e. one with $\rho_\nabla(d^2\underline{t}) = 0$, so that $\rho_\nabla([dt_i, dt_j]) = 0$, and put,

$$P^i := \rho_\nabla(dt_i), \quad J^{i,j} := \rho_\nabla(t_i dt_j - t_j dt_i).$$

A short computation then gives us,

$$[P^i, P^j] = 0, [P^i, J^{j,k}] = \delta_{i,j} P^k - \delta_{j,k} P^j$$
$$[J^{i,j}, J^{r,s}] = \delta_{j,r} J^{i,s} + \delta_{i,s} J^{j,r} - \delta_{i,r} J^{j,s} - \delta_{j,s} J^{i,r}.$$

Notice that for the Minkowski metric, this gives us the usual formulas for the commutation relations of the Lorentz Lie algebra.

As we shall see in several examples, the dynamic structure defined above is sufficiently general to serve as basis for what is usually called quantization, of the electromagnetic field. For the gravitational field, we have to do some more work.

Let us look at the last first. We then have to consider a general, non-degenerate, metric, $g = 1/2 \sum_{i=1}^{d} g_{i,j} dt_i dt_j$, and the corresponding dynamical system, $(\sigma) = ([dt_i, t_j] - g^{i,j})$. Again it is easy to see that this is not violating the relations, $[dt_i, t_j] + [t_i, dt_j] = 0$ of $Ph(C)$. Notice also that in $C(\sigma)$ we have,

$$[[dt_i, dt_j], t_k] = g^{il} \frac{\partial g^{j,p}}{\partial t_l} - g^{jk} \frac{\partial g^{i,p}}{\partial t_k},$$

meaning that the *curvature* does not commute with the action of C. Introducing $\bar{dt}_i := \sum g_{i,p} dt_p$, we find that $[\bar{dt}_i, t_j] = \delta_{i,j}$. Moreover, if we let,

$$\bar{g} := 1/2 \sum_{i=1}^{d} \bar{dt}_i^2,$$

we find, $ad(\bar{g})(t_i) = \bar{dt}_i$. Using the above, we find that there is a one-to-one correspondence between connections ∇ on a C-module V and morphisms,

$$\rho_\nabla : C(\sigma) \to End_k(V),$$

defined by,

$$\rho_\nabla(dt_i) = \sum_j g^{i,j} \nabla_{\delta_j} = \nabla_{\xi_i},$$

where $\xi_i = \sum_j g^{i,j} \delta_j$ is the dual to dt_i.

Consider now the Levi-Civita connection $\nabla_{\delta_i} = \frac{\partial}{\partial t_i} + \nabla_i$, where,

$$\nabla_i \in End_C(\Theta_C),$$

is given by the matrix formula, $\nabla_i = (\Gamma_{p,i}^q)$. Put,

$$T := 1/2 \sum_{j,k} \frac{\partial g^{j,k}}{\partial t_j} \bar{dt}_k = 1/2 \sum_{j,k,l} \frac{\partial g^{j,k}}{\partial t_j} g_{k,l} dt_l$$

and consider the inner derivation,

$$\delta := ad(g - T),$$

then after a dull computation, using the well known formula for Levi-Civita connection,

$$\frac{\partial g_{i,j}}{\partial t_k} = \sum_l (\Gamma_{k,i}^l g_{l,j} + \Gamma_{k,j}^l g_{i,l})$$

$$\frac{\partial g^{r,j}}{\partial t_k} = -\sum_l (g^{r,l} \Gamma_{k,l}^j + g^{l,j} \Gamma_{k,l}^r),$$

we obtain, in $C(\sigma)$,

$$T := -1/2 (\sum_{k,l} \Gamma_{k,l}^k dt_l + \sum_{k,p,q} g^{k,q} \Gamma_{k,q}^p g_{p,l} dt_l)$$

$$\delta(t_i) := ad(g - T)(t_i) = dt_i, \quad i = 1, ..., d.$$

Therefore we have a well-defined dynamical structure (σ), with Dirac derivation $\delta := ad(g - T)$. It is easy to see that (σ) is invariant w.r.t. isometries.

Moreover, the representation, ρ of $C(\sigma)$, defined on Θ_C, by the Levi-Civita connection, has a Hamiltonian,

$$Q := \rho(g - T) = 1/2 \sum_{i,j} g^{ij} \nabla_{\delta_i} \nabla_{\delta_j},$$

i.e. the generalized Laplace-Beltrami operator, which is also invariant w.r.t. isometries, although the proof demands some algebra. Put

$$\bar{\Gamma}^i_{p,q} := \sum_{l,r} g^{r,i} \Gamma^l_{r,p} g_{l,q},$$

then,

$$T = \sum_l T_l dt_l$$

$$T_l = -1/2 \left(\sum_j (\Gamma^j_{j,l} + \bar{\Gamma}^j_{j,l}) \right) = -1/2 (trace \nabla_l + trace \bar{\nabla}_l).$$

Since $\delta(t_i) := ad(g - T)(t_i) = dt_i$, the general force law, in $C(\sigma)$, looks like,

$$d^2 t_i = [g - T, dt_i] = -1/2 \sum_{p,q} (\bar{\Gamma}^i_{p,q} + \bar{\Gamma}^i_{q,p}) dt_p dt_q$$

$$+ 1/2 \sum_{p,q} g_{p,q} (R_{p,i} dt_q + dt_p R_{q,i})$$

$$+ [dt_i, T],$$

where, as above, $R_{i,j} = [dt_i, dt_j]$. Put,

$$\Gamma^{j,i}_p = \sum_k g^{j,k} \Gamma^i_{k,p}, \quad F_{i,j} := R_{i,j} - \sum_p (\Gamma^{j,i}_p - \Gamma^{i,j}_p) dt_p,$$

then we find,

Theorem 4.7.1 (General Force Law). *In $C(\sigma)$ we have the following force law,*

$$d^2 t_i = -\sum_{p,q} \Gamma^i_{p,q} dt_p dt_q - 1/2 \sum_{p,q} g_{p,q} (F_{i,p} dt_q + dt_p F_{i,q})$$

$$+ 1/2 \sum_{l,p,q} g_{p,q} [dt_p, (\Gamma^{i,q}_l - \Gamma^{q,i}_l)] dt_l + [dt_i, T].$$

Proof. As we have seen, the dual of dt_i is $\xi_i = \sum_l g^{i,l} \frac{\partial}{\partial t_l}$, therefore

$$[\xi_i, \xi_j] = \sum_{l,k} \left(g^{i,l} \frac{\partial g^{j,k}}{\partial t_l} \frac{\partial}{\partial t_k} - g^{j,k} \frac{\partial g^{i,l}}{\partial t_k} \frac{\partial}{\partial t_l} \right)$$

is dual to

$$\sum_{l,k,p} (g^{i,l}\frac{\partial g^{j,k}}{\partial t_l}g_{k,p}dt_p - g^{j,k}\frac{\partial g^{i,l}}{\partial t_k}g_{l,p}dt_p).$$

Using the above equations relating the derivatives of $g^{i,j}$ to the Levi-Civita connection, we find,

$$\sum_{l,k,p} (g^{i,l}\frac{\partial g^{j,k}}{\partial t_l}g_{k,p}dt_p - g^{j,k}\frac{\partial g^{i,l}}{\partial t_k}g_{l,p}dt_p) = \sum_{p}(\Gamma_p^{j,i} - \Gamma_p^{i,j})dt_p$$

where $\Gamma_p^{j,i} = \sum_k g^{j,k}\Gamma_{k,p}^i$. Let now,

$$F_{i,j} := R_{i,j} - \sum_{p}(\Gamma_p^{j,i} - \Gamma_p^{i,j})dt_p.$$

For every connection ∇ on a C-module E, given by a representation, ρ_E, we obtain,

$$\rho_E(F_{i,j}) = [\nabla_{\xi_i}, \nabla_{\xi_j}] - \nabla_{[\xi_i,\xi_j]},$$

i.e. the curvature of the connection, $F(\xi_i, \xi_j)$.

Now, plug this in the force law above, i.e. write,

$$1/2\sum_{p,q} g_{p,q}(R_{p,i}dt_q + dt_pR_{q,i}) =$$

$$1/2\sum_{p,q} g_{p,q}((R_{p,i} - \sum_{l}(\Gamma_l^{i,p} - \Gamma_l^{p,i})dt_l)dt_q + dt_p(R_{q,i} - \sum_{l}(\Gamma_l^{i,q} - \Gamma_l^{q,i})dt_l))$$

$$+ 1/2\sum_{p,q} g_{p,q}(\sum_{l}(\Gamma_l^{i,p} - \Gamma_l^{p,i})dt_l))dt_q$$

$$+ 1/2\sum_{p,q} g_{p,q}dt_p(\sum_{l}(\Gamma_l^{i,q} - \Gamma_l^{q,i})dt_l),$$

and use

$$+1/2\sum_{p} g_{p,q}(\sum_{l}(\Gamma_l^{i,p} - \Gamma_l^{p,i})dt_l))dt_q = 1/2\bar{\Gamma}_{l,q}^i dt_l dt_q - 1/2\Gamma_{q,l}^i dt_l dt_q$$

$$+1/2\sum_{q} g_{p,q}(\sum_{l}(\Gamma_l^{i,q} - \Gamma_l^{q,i})dt_p dt_l = 1/2\bar{\Gamma}_{l,p}^i dt_p dt_l - 1/2\Gamma_{p,l}^i dt_p dt_l.$$

Finally use,

$$dt_p(\Gamma_l^{i,q} - \Gamma_l^{q,i}) = (\Gamma_l^{i,q} - \Gamma_l^{q,i})dt_p + [dt_p, (\Gamma_l^{i,q} - \Gamma_l^{q,i})].$$

$\square$

We shall consider the above formula as a general Force Law, in $Ph(C)$, induced by the metric g. As explained before, this means the following: Let $\mathfrak{c}$ be the δ- stable ideal generated by this equation in $Ph^\infty(C)$. Since the force law above holds in the dynamical system defined by (σ), we obviously have $\mathfrak{c} \subset (\sigma)$, and we may hope this new dynamical system leads to a Quantum Field Theory, as defined above, with new and interesting properties. We know that this dynamical structure reduces to the generic structure for connections, i.e. for the singular cases.

Notice that this force law reduces to an equation of motion in General Relativity, in the representation-dimension 1 case, i.e. in the commutative case. More interesting is that it leads to both Lorentz force law, and to Maxwell's field-equations for Electro-Magnetism in the classical flat-space-situation, see Examples (4.12) and (4.13).

An easy calculation in $C(\sigma)$, shows that,

$$[T, dt_i] = 1/2 \sum_j T_j R_{j,i} - 1/2 \sum_{j,l} \frac{\partial T_j}{\partial t_l} g^{l,i} dt_j =: q_i.$$

But, be careful, these q_i's no longer vanish in the classical phase-space, i.e. in the commutativization of $Ph(C)$.

Now, choose a representation $\rho_E : C(\sigma) \to End_k(E)$, i.e. a connection ∇, on a C-module E. The generalized curvature $F_{i,j} =: F(\xi_i, \xi_j) \in End_C(E)$ maps to the classical one, and we observe that there is an *interaction* between the geometry, defined by the metric g and the *geometry* defined by the connection ∇. Our Force Law above will now take the form,

$$\rho_E(d^2 t_i) + \sum_{p,q} \Gamma^i_{p,q} \nabla_{\xi_p} \nabla_{\xi_q}$$

$$= 1/2 \sum_p F_{p,i} \nabla_{\delta_p} + 1/2 \sum_p \nabla_{\delta_p} F_{p,i} + 1/2 \sum_{l,q} \delta_q (\Gamma^{i,q}_l - \Gamma^{q,i}_l) \nabla_{\xi_l} + [\nabla_{\xi_i}, \rho_E(T)].$$

See also Example (4.14).

Notice also, that for the Levi-Civita connection, there is a possible relationship between this formula and the Einstein field equation. See [27], Proposition 4.2.2., p.114. If, above we assume that we are in a *geodesic reference frame*, i.e. along a geodesic γ in our space $Simp_1(C)$, then an average of the *excess-relative-acceleration*, i.e. of $d^2 t_i + \sum_{p,q} \Gamma^i_{p,q} dt_p dt_q$, evaluated in $\Theta_C|\gamma$, is proved to be given by the *Ric* tensor. But, above this relative-acceleration is, for any representation corresponding to a connection ∇, equal to

$$1/2 \sum_p F_{p,i} \nabla_{\delta_p} + 1/2 \sum_p \nabla_{\delta_p} F_{p,i} + 1/2 \sum_{l,q} \delta_q (\Gamma^{i,q}_l - \Gamma^{q,i}_l) dt_l + [\nabla_{\xi_i}, \rho_E(T)].$$

Since this excess-relative-acceleration, representing a *tidal force*, should be a measure of the inertial mass present, it is tempting to consider this force law as a generalized, quantized, Maxwell-Einstein's equation. The reference to Maxwell here is natural, since if the bundle $E = \Theta_C$ above is the tangent bundle, and we consider the connection, given by the *potential* $A = (A_1, ..., A_n), A_i \in C$, then the resulting curvature is the electro-magnetic force field. See Example (4.13) for the notion of *Charge*, and see Example (4.12), where the problem of Mass will be addressed.

In this generality, it is not really meaningful to ask for invariance of this general Force Law, w.r.t. isometries. This is linked to the fact that, in general, this force law, considered as a dynamical structure on C, may have non-singular finite-dimensional representations, and then invariance under isometries of $Simp_1(C)$ is not the proper question to pose. We shall come back to this later, but see Example (4.12) for relations to Newton and Kepler's laws.

Notice that applying ρ, corresponding to the Levi-Civita connection, the above translate into,

$$\rho(d^2 t_i) = \sum_{j=1}^{d} [Q, g^{i,j} \nabla_{\delta_j}],$$

where Q is the Laplace-Beltrami operator.

Before we turn to situations requiring a general quantum theoretical treatment, let us go back to the discussion above, about how to look at parsimony, via Lagrange functions or via dynamical systems. We claimed that the integral curves of the vector field

$$\delta = \sum_l (\xi_l \frac{\partial}{\partial t_l} - \Gamma^l \frac{\partial}{\partial \xi_l}),$$

in $Simp_1 Ph(C)$, projects onto the geodesics of the metric g in $Simp_1(C)$. These geodesics are assumed to be trajectories of *free test particles* in the geometric space $Simp_1(C)$ outfitted with the the metric g. As such they must be curves parametrized by some clock parameter τ. Since quantum field theory is assumed to model such movements, we now have two different methods to pick out such trajectories, i.e. to find the solutions $\mathbf{M}(C, k[\tau]) \subset \mathbf{F}(C, k[\tau])$. One, using dynamic systems, the force law $d^2 t_i = -\sum \Gamma^i_{p,q} dt_p dt_q$, deduced above for C, and the obvious $d^2\tau = 0$, for the free particle modeled by $B := k[\tau]$, the other using the Euler-Lagrange equations as described above.

In the first case we have the equation,

$$\delta_B(\phi) - \phi(\delta_C) = 0,$$

which evaluated at $d^2 t_i$ give us,

$$\phi(-\sum \Gamma^i_{p,q} dt_p dt_q) = (\frac{\partial}{\partial \tau})^2 (\phi_i) d\tau^2,$$

with the resulting equation,

$$\ddot{\phi}_i = -\sum \Gamma^i_{p,q} \dot{\phi}_p \dot{\phi}_q,$$

i.e. the equations for a geodesic.

In the second case, we should use the obvious derivation $\zeta = \frac{\partial}{\partial \tau}$, the corresponding representation $\rho_\zeta : Ph(k[\tau]) \to k[\tau]$, pick the Lagrangian $L := g$, and look at the resulting action and corresponding Euler-Lagrange equations. We obtain,

$$\mathbf{L} = \sum g_{p,q} \dot{\phi}_p \dot{\phi}_q$$
$$S = \int \sum g_{p,q} \dot{\phi}_p \dot{\phi}_q d\tau,$$

together with the Euler-Lagrange equations,

$$\frac{\partial g}{\partial \phi_i} - \frac{\partial}{\partial \tau}(\frac{\partial g}{\partial \dot{\phi}_i}) = 0$$

which reduces to the same equations for geodesics.

Example 4.2. With this done, let us consider some easy examples of *quantum theory*, first in dimension 1, and still in rank 1. That is, we start with the k-algebra $C = k < x >= k[x]$, and consider the classical Lagrangians,

$$L = 1/2 dx^2 - V(x) \in PhC.$$

The corresponding dynamical system σ, deduced from the Lagrange equations, as above, is given by the force law,

$$d^2 x = \frac{\partial V}{\partial x},$$

and is of order 2, so the algebra of interest is,

$$\mathbf{C}(\sigma) = PhC = k < x, dx > \simeq k < x_1, x_2 > .$$

Notice that the classical *Hamiltonian* $H := dx^2 - L$, is not an invariant, i.e. $\delta(H) \neq 0$.

Let us first compute the *particles* in rank 1 for some cases, and let us start with $V(x) = 1/2 \, x^2$, i.e. the classical oscillator. The fundamental equation of the dynamical system is,

$$\delta = [\delta] + [Q, -],$$

where, in dimension 1, the endomorphism Q obviously commutes with the actions of x_i, $i = 1, 2$. To solve the equation above, we may therefore forget about Q, so we are left with the vector fields,

$$[\delta] = \xi.$$

The space, $Simp_1(\mathbf{C}(\sigma))$, is just the ordinary phase space, $Simp_1(k[x, dx])$. Put as above, $x_1 := x, x_2 := dx$. We must solve the equations,

$$\delta(x) = [\delta](x) = [\delta](x_1)$$
$$\delta^2(x) = [\delta](dx) = [\delta](x_2)$$

We can obviously pick,

$$\delta_i = \chi_i = \frac{\partial}{\partial x_i},$$

so we must have

$$[\delta] = \xi_1 \frac{\partial}{\partial x_1} + \xi_2 \frac{\partial}{\partial x_2}.$$

In the case of the potential, $V = 1/2x^2$, we get the equations,

$$x_2 = [\delta](x) = [\delta](x_1) = \xi_1$$
$$x_1 = [\delta](dx) = [\delta](x_2) = \xi_2$$

Therefore the fundamental vector field is,

$$\xi = x_2 \frac{\partial}{\partial x_1} + x_1 \frac{\partial}{\partial x_2}$$

i.e. we find hyperbolic motions in the phase space, with general solutions,

$$x = x_1 = r\cosh(t + c), \ dx = x_2 = r\sinh(t + c)$$

which is what we expected.

In the case of the oscillator, $V = -1/2x^2$, we get the equations,

$$x_2 = [\delta](x) = [\delta](x_1) = \xi_1$$
$$-x_1 = [\delta](dx) = [\delta](x_2) = \xi_2$$

Therefore the fundamental vector field is,

$$\xi = x_2 \frac{\partial}{\partial x_1} - x_1 \frac{\partial}{\partial x_2}$$

i.e. we find circular motions in the phase space, with general solutions,

$$\gamma : x = x_1 = r\cos(t + c), \ dx = x_2 = -r\sin(t + c),$$

which is also what we expected.

Consider now the versal family restricted to γ,

$$\tilde{\rho}_\gamma : k < x, dx > \longrightarrow End_\gamma(\tilde{V}|\gamma),$$

and a state $\psi(t) \in \tilde{V}|\gamma$. If Q, restricted to γ, is multiplication by $\kappa(t)$, (in physics, one usually puts $\kappa(t) = \imath\kappa$), then the Schrödinger equation becomes,

$$\frac{\partial}{\partial t}\psi = \kappa(t)\psi$$

so that we should have,

$$\psi(t) = exp(\int_\gamma^t \kappa).$$

This will turn out much nicer if we extend the action of $k < x_1, x_2 >$ to $\tilde{V}_C$, and put Q, restricted to γ, equal to multiplication by $\imath\kappa$. Then we find the reasonable result,

$$\psi(t) = exp(\imath \int_\gamma^t \kappa).$$

See again [19].

In the repulsive, resp. attractive, Newtonian case, with $V = \pm 1/x$, we find,

$$x_2 = [\delta](x) = [\delta](x_1) = \xi_1$$
$$\epsilon(1/x_1^2) = [\delta](dx) = [\delta](x_2) = \xi_2, \quad \epsilon = +, -.$$

Therefore the fundamental vector field is,

$$\xi = x_2 \frac{\partial}{\partial x_1} + \epsilon(1/x_1^2)\frac{\partial}{\partial x_2}$$

with the classical solution,

$$x = \epsilon(9/2)t^{2/3}.$$

In higher dimensions, say in the case of our toy model H, of the Introduction, this rank 1 theory reduces to the wave-mechanics of de Broglie. Recall that there is a natural action of the Lie group, $U(1)$ on B_o, and therefore a natural complex structure on the tangent space $T_{\underline{H}}$. Consider the trivial versal family,

$$Ph(H) \to End_{C(1)}(C(1), C(1)),$$

where we may assume $C(1)$ is a complex vector space. Any order 2 dynamical structure defined on H, will induce a vector field in $\underline{C}(1) = \underline{Ph(H)}$, which in the case of the Levi-Civita connection, considered as force law, makes the integral curves geodesics. From this the Klein-Gordon equation follows in a natural way, and in this context we may also discuss interference and diffraction of light, see [20].

Example 4.3. Now let us go back to the case of $A = k < x_1, x_2 >$, the free non-commutative k-algebra on two symbols, and the rank $n = 2$, see (3.3). We found,

$$C(2) \simeq k[t_1, t_2, t_3, t_4, t_5].$$

locally, in a Zariski neighborhood of the origin. The versal family $\tilde{V}$, is defined by the actions of x_1, x_2, given by,

$$X_1 := \begin{pmatrix} 0 & 1+t_3 \\ t_5 & t_4 \end{pmatrix}, \ X_2 := \begin{pmatrix} t_1 & t_2 \\ 1+t_3 & 0 \end{pmatrix}.$$

The Formanek center, in this case, is cut out by the single equation:

$$f := det[X_1, X_2] = -((1+t_3)^2 - t_2 t_5)^2 + (t_1(1+t_3) + t_2 t_4)(t_4(1+t_3) + t_1 t_5).$$

and

$$tr X_1 = t_4, \ tr X_2 = t_1,$$
$$det X_1 = -t_5 - t_3 t_5, \ det X_2 = -t_2 - t_2 t_3,$$
$$tr(X_1 X_2) = (1+t_3)^2 + t_2 t_5,$$

so the *trace ring* of this family is ,

$$k[t_1, t_2 + t_2 t_3, 1 + 2t_3 + t_3^2 + t_2 t_5, t_4, t_5 + t_3 t_5] =: k[u_1, u_2, u_3, u_4, u_5],$$

with,

$$u_1 = t_1, \ u_2 = (1+t_3)t_2, \ u_3 = (1+t_3)^2 + t_2 t_5, \ u_4 = t_4, \ u_5 = (1+t_3)t_5,$$

and $f = -u_3^2 + 4u_2 u_5 + u_1 u_3 u_4 + u_1^2 u_5 + u_2 u_4^2$. Moreover, $k[\underline{t}]$ is algebraic over $k[\underline{u}]$, with discriminant, $\Delta := 4u_2 u_5(u_3^2 - 4u_2 u_5) = 4(1+t_3)^2 t_2 t_5((1+t_3)^2 - t_2 t_5)^2$, and there is an étale covering,

$$\mathbf{A}^5 - V(\Delta) \to Simp_2(A) - V(\Delta).$$

Notice that if we put $t_1 = t_4 = 0$, then f divides Δ.

Example 4.4. Quantum field theory for the oscillator, given by the Lagrangian, $L = 1/2dx^2 - 1/2x^2$, and with the force law, $d^2 x = x$, in rank 2, is more difficult. Above we have found a (partial) versal family of $Simp_2(Ph\ k[x])$, over the versal base space $C(2) = k[t_1, ..., t_5]$, given by,

$$x = \begin{pmatrix} 0 & 1+t_3 \\ t_5 & t_4 \end{pmatrix}, dx = \begin{pmatrix} t_1 & t_2 \\ 1+t_3 & 0 \end{pmatrix}.$$

The fundamental vector fields will have the form,

$$[\delta] = \sum \xi_i \delta_i, \quad \xi = \sum \xi_i \frac{\partial}{\partial t_i},$$

with 5 unknowns, $\xi_i, i = 1, 2, .., 5$. Moreover,

$$Q = \begin{pmatrix} q_{1,1} & q_{1,2} \\ q_{2,1} & q_{2,2} \end{pmatrix},$$

with 4 unknowns $q_{i,j}, i = 1, 2, j = 1, 2$. Now, recall that Q can only be determined up to a central element from $M_2(C)$, i.e. we have 8 essential unknowns, $\xi_i, i = 1, 2, 3, 4, 5$ and $(q_{1,1} - q_{2,2}), q_{1,2}, q_{2,1}$ in the two matrix equations,

$$\delta(x) = dx = [\delta](x) + [Q, x]$$
$$\delta^2(x) = x = [\delta](dx) + [Q, dx]$$

On the right hand side of the equations we have the terms,

$$[\delta](x) = \sum \xi_i \delta_i \left(\begin{pmatrix} 0 & 1 + t_3 \\ t_5 & t_4 \end{pmatrix} \right) = \begin{pmatrix} 0 & \xi_3 \\ \xi_5 & \xi_4 \end{pmatrix}$$

$$[\delta](dx) = \sum \xi_i \delta_i \left(\begin{pmatrix} t_1 & t_2 \\ 1 + t_3 & 0 \end{pmatrix} \right) = \begin{pmatrix} \xi_1 & \xi_2 \\ \xi_3 & 0 \end{pmatrix}$$

and the terms,

$$[Q, x] = \begin{pmatrix} t_5 q_{1,2} - (1 + t_3) q_{2,1} & (1 + t_3) q_{1,1} + t_4 q_{1,2} - (1 + t_3) q_{2,2} \\ t_5 q_{2,2} - t_5 q_{1,1} - t_4 q_{2,1} & (1 + t_3) q_{2,1} - t_5 q_{1,2} \end{pmatrix}$$

$$[Q, dx] = \begin{pmatrix} (1 + t_3) q_{1,2} - t_2 q_{2,1} & t_2 q_{1,1} - t_1 q_{1,2} - t_2 q_{2,2} \\ t_1 q_{2,1} + (1 + t_3) q_{2,2} - (1 + t_3) q_{1,1} & t_2 q_{2,1} - (1 + t_3) q_{1,2} \end{pmatrix},$$

and on the left side, we have,

$$\delta(x) = dx = \begin{pmatrix} t_1 & t_2 \\ 1 + t_3 & 0 \end{pmatrix}$$

$$\delta^2(x) = x = \pm \begin{pmatrix} 0 & 1 + t_3 \\ t_5 & t_4 \end{pmatrix}.$$

Writing up the matrix for the corresponding linear equation, we find that the determinant of the 8×8 matrix turns out to be easily computed, it is,

$$D = 2(1 + t_3)(t_2 t_5 - (1 + t_3)^2).$$

Notice that D is a divisor in the discriminant, $\Delta = 4(1+t_3)^2 t_2 t_5 ((1+t_3)^2 - t_2 t_5)^2$, see (3.5). Moreover we find,

$$(q_{1,1} - q_{2,2}) = D^{-1}(-(1+t_3)(t_1^2 + t_4^2) + (t_2 - t_5)(t_2 t_5 - (1+t_3)^2 - t_1 t_4))$$
$$q_{1,2} = D^{-1}(2(1+t_3)(t_1 t_2 + (1+t_3)t_4)$$
$$q_{2,1} = D^{-1}(2(1+t_3)(t_4 t_5 + t_1(1+t_3))$$
$$\xi_1 = t_2 q_{2,1} - (1+t_3)q_{1,2}$$
$$\xi_2 = -t_2(q_{1,1} - q_{2,2}) + t_1 q_{1,2} + (1+t_3)$$
$$\xi_3 = (1+t_3)(q_{1,1} - q_{2,2}) + t_1 q_{2,1} + t_5$$
$$\xi_4 = t_5 q_{1,2} - (1+t_3)q_{2,1}$$
$$\xi_5 = t_5(q_{1,1} - q_{2,2}) + t_4 q_{2,1} + (1+t_3)$$

See that $\xi_1 = \xi_4 = 0$ imply,

$$((1+t_3)^2 - t_2 t_5)q_{1,2} = ((1+t_3)^2 - t_2 t_5)q_{2,1} = 0,$$

and, since we assume that $\Delta \neq 0$, therefore, $((1+t_3)^2 - t_2 t_5) \neq 0$, and so $q_{1,2} = q_{2,1} = 0$, this also implies that $t_1 = t_4 = 0$. Therefore the singularities of ξ are given, by,

$$t_2 = -(1+t_3), \quad t_5 = +(1+t_3),$$

or, up to isomorphisms, uniquely, by the representation,

$$x = \begin{pmatrix} 0 & 1 \\ 1 & 0 \end{pmatrix}$$

$$dx = \begin{pmatrix} 0 & -1 \\ 1 & 0 \end{pmatrix}$$

$$Q = \begin{pmatrix} q_{1,1} & 0 \\ 0 & q_{1,1}+1 \end{pmatrix}.$$

corresponding to $t_1 = 0, t_2 = -1, t_3 = 0, t_4 = 0, t_5 = 1$. Notice that in this case we find, in all ranks, that $f_\hbar := \rho(x + dx)$, is an eigenvector for $[Q, -]$ with $f_{-\hbar} = \rho(x - dx)$ so that $N = f_{-\hbar} f_\hbar$ is the *quantum counting operator*.

Let us pause a little, to compute the gradient of the action, $S = Tr(\mathbf{L})$. Since $L = 1/2 dx^2 + 1/2x^2$, this is easy, and we find,

$$S = 1/2(t_1^2 + 2t_2(1+t_3) + 2t_5(1+t_3) + t_4^2), \nabla S = (t_1, (1+t_3), (t_2+t_5), t_4, (1+t_3)),$$

which, clearly is different from the vector field ξ above, see the Introduction. But the singularities, obtained by solving $\nabla S = 0$, for,

$$x = \begin{pmatrix} 0 & 1+t_3 \\ t_5 & t_4 \end{pmatrix}, dx = \begin{pmatrix} t_1 & t_2 \\ 1+t_3 & 0 \end{pmatrix}$$

gives us,

$$x = \begin{pmatrix} 0 & 0 \\ -t_2 & 0 \end{pmatrix}, dx = \begin{pmatrix} 0 & t_2 \\ 0 & 0 \end{pmatrix}$$

which is isomorphic to the singularitity for ξ, after the coordinate change, $a^+ := 1/2(x+dx), a := 1/2(x-dx)$, with the same Hamiltonian Q. Notice, however, that even though the Formanek center f, is non-vanishing, our family is not good at this point. Since $1 + t_3 = 0$, the discriminant $\Delta = 0$, and so our family is not étale at this point.

Now, to find the integral curves of the vector field ξ, we must solve the obvious system of differential equations, $\frac{\partial t_i}{\partial \tau} = \xi_i, i = 1, .., 5$. It turns out that we are mostly interested in the solutions for which there exists singular point, corresponding to $t_1 = t_4 = 0$. If they exist they look like,

$$\frac{\partial t_1}{\partial \tau} = \xi_1 = 0$$

$$\frac{\partial t_2}{\partial \tau} = \xi_2 = -t_2(t_2 - t_5)(2 + 2t_3)^{-1} + (1 + t_3)$$

$$\frac{\partial t_3}{\partial \tau} = \xi_3 = 1/2(t_2 - t_5) + t_5$$

$$\frac{\partial t_4}{\partial \tau} = \xi_4 = 0$$

$$\frac{\partial t_5}{\partial \tau} = \xi_5 = t_5(t_2 - t_5)(2 + 2t_3)^{-1} + (1 + t_3).$$

And these equations are obviously consistent with the conditions $t_1 = t_4 = 0$.

Introducing new variables,

$$y_1 = (t_2 - t_5)$$
$$y_2 = (t_2 + t_5)$$
$$y_3 = (2 + 2t_3)$$

so that,

$$t_2 = 1/2 \, (y_2 + y_1)$$
$$t_5 = 1/2 \, (y_2 - y_1)$$
$$t_3 = 1/2 \, y_3 - 1.$$

things look nicer. We find,

$$\xi_2 = -1/2 \, (y_1 + y_2) + 1/2 \, y_3$$
$$\xi_3 = 1/2 \, y_2$$
$$\xi_5 = 1/2 \, (y_2 - y_1)y_1 y_3^{-1}$$

In the new coordinates the system of equations above reduces to,

$$y_1 \frac{\partial y_1}{\partial \tau} - y_2 \frac{\partial y_2}{\partial \tau} + y_3 \frac{\partial y_3}{\partial \tau} = 0$$

$$y_1^{-1} \frac{\partial y_1}{\partial \tau} + y_3^{-1} \frac{\partial y_3}{\partial \tau} = 0.$$

The integral curves are therefore intersections of the form,

$$C(c_1, c_2) := V(y_1^2 - y_2^2 + y_3^2 = c_1) \cap V(y_1 y_3 = c_2).$$

Moreover, the stratum at infinity, given by $f = 0$, where f is the Formanek center, is now easily computed, in terms of the new coordinates it is given as,

$$f = -1/16(y_1^2 - y_2^2 + y_3^2)^2$$

This shows that a particle corresponding to an integral curve $\gamma := C(c_1, c_2)$, with $c_1 \neq 0$ *lives eternally*, as it should. Its completion does not intersect the Formanek center, the stratum at infinity.

An easy calculation gives us, see Example (4.2),

$$16(y_1^2 - y_2^2 + y_3^2)^2 = -u_3^2 + 4u_2 u_5$$

$$y_1 y_3 = 2(u_2 - u_5),$$

where the u-coordinates are those of the trace ring, see Example (4.2). The integral curves of the harmonic oscillator will be therefore be plane conic curves in the part of $Simp_2(Phk[x])$, where $\Delta \neq 0$, $u_1 = u_4 = 0$, given by,

$$u_3^2 - 4u_2 u_5 = c_3, \ (u_2 - u_5) = c_4.$$

Here $c_3 \neq 0, c_4$ are constants. Notice also that our special point, the singularity for ξ, given by $y_1 = -2, y_2 = 0, y_3 = 2$, sits on the curve defined by $c_1 = 8, c_2 = -4$, corresponding to $c_3 = 32$, $c_4 = -2$.

In the new, y-coordinates, the versal family of $Simp_2(Ph \, k[x])$, lifted to $U(2)$, and restricted to $t_1 = t_2 = 0$, is given by,

$$x = \begin{pmatrix} 0 & 1/2y_3 \\ 1/2(y_2 - y_1) & 0 \end{pmatrix}, dx = \begin{pmatrix} 0 & 1/2(y_1 + y_2) \\ 1/2y_3 & 0 \end{pmatrix}.$$

Moreover along the curve γ, defined by $c_1 = 8, c_2 = -4$, which is given by the equations,

$$y_3 = -4y_1^{-1}, \ y_2^2 = y_1^2 + 16y_1^{-2} - 8 = (y_1^2 - 4)^2 y_1^{-2},$$

the vectorfield ξ is given by,

$$\xi = -1/4(y_1 + 2)(y_1 - 2)y_1 \frac{\partial}{\partial y_1}$$

or,

$$\xi = 3/4(y_1 + 2)(y_1 - 2)y_1\frac{\partial}{\partial y_1},$$

depending on which root we choose for y_2 above. The corresponding *time* along γ, is then given as, $\tau = -log(y_1) + 1/2log(y_1 + 2) + 1/2log(y_1 - 2)$, respectively $\tau = 1/3log(y_1) - 1/6log(y_1 + 2) - 1/6log(y_1 - 2)$, both with a singularity at $y_1 = -2$, $y_1 = 2$, corresponding to the same unique singularity of ξ, in $Simp_2(Ph(k[x])$. This shows that to reach the singularity, from outside, would take infinite time.

The versal family is not defined at $y_1 = 0$, see above.

Example 4.5. (i) We shall not treat oscillators in rank≥ 3, in general, but only look at the singularities, in all ranks. This is all well known in physics, see [2], section 16, although in most books in physics, it is treated rather formally, in relation with the *second quantification* and the introduction of Fock-spaces, and their associated representations of the algebra of observables. We shall see that this second quantification is a natural quotient of the algebra of observables PhC, in line with the general philosophy of this paper. Although we may work in a very general setting, we shall, as above, restrict our attention to the classical oscillator $L = 1/2dx^2 - 1/2x^2$), in dimension 1.

As above we find,

$$d^2x = x$$

and the Dirac derivation has therefore,

$$a_+ := 1/2(x + dx), \quad a_- := 1/2(x - dx)$$

as eigenvectors, with eigenvalues 1 and -1 respectively. Since $Ph(C) = k < x, dx >$ is generated by the elements $a_+ := 1/2(x + dx)$, $a_- := 1/2(x - dx)$, it is clear that Planck's constant $\hbar = 1$. Notice also that the classical Hamiltonian is given by,

$$Q := dx^2 - L = 2a_+a_-.$$

Using the methode above it is easy to see that for any rank $n = dimV$, a singular point $v \in Simp_n(PhC)$ corresponds to a $k < x, dx >$-module V, with x and dx acting as endomorphisms $X, dX \in End_k(V)$ for which there exists an endomorphism, the Hamiltonian, $Q \in End_k(V)$ with,

$$dX := \rho(dx) = [Q, \rho(x)] =: [Q, X]$$
$$X = \rho(d^2x) = [Q, \rho(dx)] =: [Q, dX]$$

Let ψ_0 be any eigenvector for Q with eigenvalue κ_0. Since V is simple, the family $\{a_+^m a_-^n (\psi_0)\}$ must generate V. Moreover, if $a_+^m a_-^n (\psi_0) \neq 0$, we know it must be an eigenvector for Q, with eigenvalue $\kappa_0 + (m - n)$. We can, by adding $\lambda \mathbf{1}$ to Q, assume that there is a basis for V of eigenvectors for Q, with eigenvalues of this form. This means that Q can be assumed to have the form,

$$
Q = \begin{pmatrix}
\kappa_0 & 0 & 0 & 0 \dots & 0 \\
0 & \kappa_0 + \lambda_1 & 0 & 0 \dots & 0 \\
0 & 0 & \kappa_0 + \lambda_2 & 0 \dots & 0 \\
\cdot & \cdot & \cdot & \cdot \dots & 0 \\
0 & 0 & 0 & \dots 0 & \kappa_0 + \lambda_{n-1}
\end{pmatrix},
$$

where $0 \leq \lambda_1 \leq \lambda_2 \leq \dots \leq \lambda_{n-1}$ are all integers. Moreover, since V is simple, and $[Q, a_+] = a_+$, $[Q, a_-] = -a_-$, an easy computation shows that,

$$
a_+ = \begin{pmatrix}
0 & 0 & 0 & 0 & \dots & 0 \\
a_{2,1} & 0 & 0 & 0 & \dots & 0 \\
0 & a_{3,2} & 0 & 0 & \dots & 0 \\
\cdot & \cdot & \cdot & & \dots & 0 \\
0 & 0 & 0 & \dots a_{n,n-1} & 0
\end{pmatrix},
$$

$$
a_- = \begin{pmatrix}
0 & a_{1,2} & 0 & 0 & \dots & 0 \\
0 & 0 & a_{2,3} & 0 & \dots & 0 \\
0 & 0 & 0 & 0 & \dots & 0 \\
\cdot & \cdot & \cdot & \cdot & \dots a_{n-1,n} \\
0 & 0 & 0 & \dots 0 & 0
\end{pmatrix},
$$

where all $a_{i,i-1}, a_{i,i+1} \neq 0$. We also find,

$[a_+, a_-]$

$$
= \begin{pmatrix}
-a_{1,2}a_{2,1} & 0 & 0 & 0 \dots & 0 \\
0 & a_{2,1}a_{1,2} - a_{2,3}a_{3,2} & 0 & 0 \dots & 0 \\
0 & 0 & a_{3,2}a_{2,3} - a_{3,4}a_{4,3} & 0 \dots & 0 \\
\cdot & \cdot & \cdot & \cdot \dots & 0 \\
0 & 0 & 0 & \dots 0 & a_{n,n-1}a_{n-1,n}
\end{pmatrix}
$$

obviously with vanishing trace.

Now to have the classical formulas, see ([2], p.377-380), we just have to impose the condition that a_+ and a_- be *conjugate operators*, i.e. that

$$
[a_+, a_-] = \begin{pmatrix}
-1 & 0 & 0 & \dots & 0 \\
0 & -1 & 0 & \dots & 0 \\
0 & 0 & -1 & \dots & 0 \\
\cdot & \cdot & \cdot & \dots & 0 \\
0 & 0 & 0 & \dots & (n-1)
\end{pmatrix}.
$$

Then, introducing a base change, corresponding to an inner automorphism defined by a diagonal matrix, we find that we may assume $a_{i,i+1} = a_{i+1,i}$. It follows that,

$$X = \begin{pmatrix} 0 & \sqrt{1} & 0 & 0 & \cdots & & 0 \\ \sqrt{1} & 0 & \sqrt{2} & 0 & \cdots & & 0 \\ 0 & \sqrt{2} & 0 & \sqrt{3} & \cdots & & 0 \\ \cdot & \cdot & \cdot & \cdot & \cdots & \sqrt{(n-1)} & \\ 0 & 0 & 0 & \cdots & \sqrt{(n-1)} & & 0 \end{pmatrix}$$

$$dX = \begin{pmatrix} 0 & -\sqrt{1} & 0 & 0 & \cdots & & 0 \\ \sqrt{1} & 0 & -\sqrt{2} & 0 & \cdots & & 0 \\ 0 & \sqrt{2} & 0 & -\sqrt{3} & \cdots & & 0 \\ \cdot & \cdot & \cdot & \cdot & \cdots & -\sqrt{(n-1)} & \\ 0 & 0 & 0 & \cdots & \sqrt{(n-1)} & & 0 \end{pmatrix}$$

with associated Hamiltonian,

$$Q = \begin{pmatrix} 1/2 & 0 & 0 & 0 & \cdots & 0 \\ 0 & 3/2 & 0 & 0 & \cdots & 0 \\ 0 & 0 & 5/2 & 0 & \cdots & 0 \\ \cdot & \cdot & \cdot & \cdot & \cdots & 0 \\ 0 & 0 & 0 & \cdots & 0 & (2n-1)/2 \end{pmatrix}.$$

Clearly, we cannot impose, $[a_-, a_+] = \mathbf{1}$, in finite rank. If, however, we let $n = dim_k V$ tend to ∞, then we find exactly the classical formulas for the oscillator as in the second quantification, see the reference above. In particular it follows that $[a_-, a_+] = 1$ is the only relation between the operators a_- and a_+ in this classical limit representation.

On the basis of the examples above, in particular Example (4.4), it is tempting to conjecture that all integral curves of ξ are intersections of hypersurfaces of $Spec(C(n))$, of the form $Tr\xi(\tilde{\rho}(\theta)) = const.$. However, this is not true, as we can see by going back to Example (4.3). Here we have

$$A = k[x], \quad \mathbf{A}(\sigma) = PhA = k < x, dx >= k < x, y >, \quad y = dx, \quad \delta = y\frac{\partial}{\partial x} + x\frac{\partial}{\partial y}.$$

There are only two obvious invariants, $\theta_1 = x^2 - y^2$, i.e. the Hamiltonian, and $\theta_2 = xy - yx$. Moreover the universal family on $C(2) = k[t_1, .., t_5]$, is given by,

$$\tilde{\rho}(x) = \begin{pmatrix} 0 & 1+t_3 \\ t_5 & t_4 \end{pmatrix}, \tilde{\rho}(y) = \begin{pmatrix} t_1 & t_2 \\ 1+t_3 & 0 \end{pmatrix}.$$

We find, see (3.5), that the invariants expressed in the coordinates $(u_1, ..., u_5)$, looks like,

$$trace(\tilde{\rho}(\theta_1)) = -u_1 - 2u_2 + u_4 + 2u_5$$
$$det(\tilde{\rho}(\theta_1)) = (u_5 - u_2 - u_1^2)(u_5 - u_2 + u_4^2) - u_4^2 u_5 + u_1 u_3 u_4 - u_1^2 u_2.$$
$$det(\tilde{\rho}(\theta_2)) = -u_3^2 + 4u_2 u_5 + u_1 u_3 u_4 + u_1^2 u_5 + u_2 u_4^2$$
$$det(\tilde{\rho}(\theta_1)\tilde{\rho}(\theta_2)) = 0.$$

Recall from above that,

$$u_1 = t_1, \ u_2 = (1+t_3)t_2, \ u_3 = (1+t_3)^2 + t_2 t_5, \ u_4 = t_4, \ u_5 = (1+t_3)t_5,$$

and,

$$\xi_2 = -1/2 \ (y_1 + y_2) + 1/2 \ y_3$$
$$\xi_3 = 1/2 \ y_2$$
$$\xi_5 = 1/2 \ (y_2 - y_1)y_1 y_3^{-1}.$$

If we put $t_1 = t_4 = 0$, we find the result of Example (4.3), namely $Tr(\tilde{\rho}(\theta_1)) = y_1 y_3 = 2(u_5 - u_2)$, $det(\tilde{\rho}(\theta_1)) = 1/4(y_1^2 y_3^2) = (u_5 - u_2)^2$, $det(\tilde{\rho}(\theta_2)) = -1/16(y_1^2 - y_2^2 + y_3^2)^2 = -u_3^2 + 4u_2 u_5$. However, the fact that $det(\tilde{\rho}(\theta_1)\tilde{\rho}(\theta_2)) = 0$ indicates that there are non-algebraic integral curves sitting on an algebraic surface of $\mathbf{A}^5$. This is related to the problem of hyperbolicity of complex algebraic surfaces. In fact, we see that any integral curve of $\xi = [\delta]$ is sitting on an algebraic surface, and we may find one for which ξ have no singularities. Is the integral curve algebraic, or may it be dense on the surface, in the Zariski topology? Exact conditions on algebraic surfaces for being hyperbolic seems not to be known. Notice moreover that the *non-commutative invariant* θ_2 is essential in the integration of ξ in this case. Notice also that when $A = k[x_1, x_2, x_3]$, and if the Lagrangian $L = 1/2(dx_1^2) + 1/2(dx_2^2) + 1/2(dx_3^2) + U$, has a potential U, such that $\frac{\partial U}{\partial x}_i x_j = \frac{\partial U}{\partial x}_j x_i$, i.e. concerns a *central force*, then the *angular momenta* $L_{i,j} := x_i dx_j - x_j dx_i$, are constants, i.e. $\delta(L_{i,j}) = 0$, in rank 1, which of course have the classical consequences one knows. Combining this with the representations discussed in the Example (2.1), (iii), we find interesting results, see next section.

Emmy Noethers theorem is, in this context, reduced to the following observation. Suppose a non-trivial derivation, ξ of $\mathbf{A}(\sigma) \simeq Ph(A)$ leaves the dynamical structure of the versal family $\tilde{\rho}$-invariant, i.e. suppose,

$$\tilde{\rho}([\xi, \delta]) = 0.$$

Let δ and ξ correspond, via Theorem (4.2.1), to the derivations $[\delta]$, resp. $[\xi]$ of $C(n)$, and to the Hamiltonians, Q, resp. Q_ξ. For all $a \in \mathbf{A}(\sigma)$ we must have,

$$\tilde\rho([[\xi,\delta],a]) = [[\xi],[\delta]](\tilde\rho(a)) + [[Q_\xi,Q],\tilde\rho(a)].$$

In the singular case, i.e. when $[\delta] = 0$, this proves that Q_ξ is a constant of the theory.

Example 4.6. We might try to find functions, or formal power series, $[n] \in k[[\tau]]$ such that the representation,

$$x(n) = \begin{pmatrix} 0 & \sqrt{[1]} & 0 & 0 & \cdots & & 0 \\ \sqrt{[1]} & 0 & \sqrt{[2]} & 0 & \cdots & & 0 \\ 0 & \sqrt{[2]} & 0 & \sqrt{[3]} & \cdots & & 0 \\ \cdot & \cdot & \cdot & \cdot & \cdots & & \sqrt{[(n-1)]} \\ 0 & 0 & 0 & \cdots & \sqrt{[(n-1)]} & & 0 \end{pmatrix}$$

$$dx(n) = \begin{pmatrix} 0 & -\sqrt{[1]} & 0 & 0 & \cdots & & 0 \\ \sqrt{[1]} & 0 & -\sqrt{[2]} & 0 & \cdots & & 0 \\ 0 & \sqrt{[2]} & 0 & -\sqrt{[3]} & \cdots & & 0 \\ \cdot & \cdot & \cdot & \cdot & \cdots & & -\sqrt{[(n-1)]} \\ 0 & 0 & 0 & \cdots & \sqrt{[(n-1)]} & & 0 \end{pmatrix}$$

with associated Hamiltonian,

$$Q = \begin{pmatrix} 1/2+[0] & 0 & 0 & 0 & \cdots & 0 \\ 0 & 1/2+[1] & 0 & 0 & \cdots & 0 \\ 0 & 0 & 1/2+[2] & 0 & \cdots & 0 \\ \cdot & \cdot & \cdot & & \cdots & 0 \\ 0 & 0 & 0 & \cdots & 0 & 1/2+[n-1] \end{pmatrix}$$

satisfiy the fundamental dynamical equation,

$$\delta = [\delta] + [Q, -].$$

We may, of course choose $[\delta] := \frac{\partial}{\partial\tau}$ as the generator of the vector fields on the τ-line. We find the following system of differential equations,

$$\frac{\partial}{\partial\tau} f_n + (f_n^2 - f_{n-1}^2) f_n = f_n$$

$$\frac{\partial}{\partial\tau} f_n + (-f_n^2 - f_{n-1}^2) f_n = -f_n$$

where $f_n := \sqrt{[n]}$, and with boundary conditions,

$$f_n(0)^2 = n.$$

These equations immediately lead to $\delta(f_n) = 0$, so to constant $f_n's$, and therefore proves that the curve in $Simp_n(PhC)$ defined by the family $\{x(n), dx(n)\}$ is transversal to the fundamental vector field ξ. The introduction of the (p,q) commutators, and their treatment in physics, makes it possible to treat the fermions and the bosons in a common structure. Letting the parameter q in the above family slide from 1 to -1, the q-commutator $[-, -]_q$ changes from the ordinary Lie product to the Jordan product. The computation above shows that this change takes place transversal to time, i.e. instantanously!

Example 4.7. For the harmonic oscillator in dimension $n = 2$ we have $A = k[x_1, x_2]$, and, $Ph(A) = k < x_1, x_2, dx_1, dx_2 > /([x_1, x_2], [x_1, dx_2] - [x_2, dx_1])$, and,

$$\mathbf{A}(\sigma) = k < x_1, x_2, dx_1, dx_2 > /([x_1, x_2], [x_1, dx_2] - [x_2, dx_1], [dx_1, dx_2]).$$

Moreover, in rank 2 we find a simple representation of $\mathbf{A}(\sigma)$, given by,

$$X_1 = \begin{pmatrix} 1 & 0 \\ 0 & 0 \end{pmatrix}, X_2 = \begin{pmatrix} 0 & 0 \\ 0 & 1 \end{pmatrix}$$

$$dX_1 = \begin{pmatrix} 0 & -1 \\ 1 & 0 \end{pmatrix}, dX_2 = \begin{pmatrix} 0 & 1 \\ -1 & 0 \end{pmatrix}$$

with,

$$[X_1, dX_1] = [X_2, dX_2] = \begin{pmatrix} 0 & -1 \\ -1 & 0 \end{pmatrix}.$$

Example 4.8. For the quartic *anharmonic* oscillator, given by $L = 1/2 \, dx^2 - 1/4 \, \alpha x^4$ we may easily compute the rank 2 and 3 versal families. In rank 2 we find that there is a one-dimensional singular family of dimension 2 simple modules, with,

$$X = \begin{pmatrix} 0 & \alpha t^3 \\ t & 0 \end{pmatrix}, dX = \begin{pmatrix} 0 & -\alpha^2 t^5 \\ \alpha t^3 & 0 \end{pmatrix}, Q = X = \begin{pmatrix} 0 & 0 \\ 0 & \alpha t^2 \end{pmatrix}.$$

In rank 3 we find that there are no simple singular module with corresponding diagonal Hamiltonian. This may be one reason why the energy levels of the quartic anharmonic oscillator is not known to the physicists.

Example 4.9. Now, let us consider the infinite rank case. In particular we may consider the representation given in the above example, when $n = dim_k V$ tends to ∞. Notice that this is given as the limit case of the singular simple representation of the classical oscillator in dimension n, with an obvious conjugation condition imposed. For $k = \mathbf{R}$, we have a real Planck's constant which we obviously may assume equal to $\hbar = 1$.

Moreover, we now have, $[a_+, a_-] = 1$, and we have a representation of $Ph(C)$ onto the algebra $\mathbf{F}$, generated by $\{a_+, a_-\}$. Notice that in each finite rank, this algebra generate the whole $End_k(V)$. The commutation relations is given by a classical formula,

$$a_-^m a_+^n = a_+^n a_-^m + mn\, a_-^{n-1} a_+^{m-1} + 1/2!\, m(m-1)n(n-1)\, a_+^{n-2} a_-^{m-2}$$
$$+ 1/3!\, m(m-1)(m-2)n(n-1)(n-2)\, a_+^{n-3} a_-^{m-3} + \ldots$$

and the Lie algebra $\mathfrak{f}$, of derivations of $\mathbf{F}$ are easily seen to be generated by the derivations $\{\delta_{p,q}\}_{p,q}$, defined as,

$$\delta_{p,q}(a_+) = a_+^p a_-^q, \quad \delta_{p,q}(a_-) = -p/(q+1)a_+^{p-1} a_-^{q+1}.$$

If we put, for $m, n \geq 0$

$$\chi_{m,n} := \delta_{m+1,n}, \quad \chi_m := \chi_{m,0}$$

then we find the Witt-algebra, with the classical relations,

$$[\chi_m, \chi_n] = (n - m)\chi_{m+n}.$$

Moreover we find,

$$[\chi_0, \chi_{m,n}] = (m - n)\chi_{m,n} =: deg(\chi_{m,n})\chi_{m,n}.$$

Clearly the Lie algebra $Der_k(\mathbf{F})$ has an ascending filtration with respect to the degree, deg, defined above, and it is easy to see that the corresponding graded Lie algebra $\mathfrak{g} := gr(Der_k(\mathbf{F}))$ has the following products,

$$[\chi_{p,q}, \chi_{r,s}] = (r - p + (s+1)^{-1}(r+1)q - (q+1)^{-1}(p+1)s)\chi_{p+r,q+s}.$$

In particular the degree zero component of $\mathfrak{g}$ is Abelian.

Example 4.10. Finally let $C := \mathbf{R}[x]$, and let $\mathbf{C} := C \otimes_{\mathbf{R}} \mathbf{C}$, and consider some representation on $V = \mathbf{C}$ of $Ph(C) = \mathbf{R} < x, dx >$. Clearly,

$$Ext_C^1(V, V) = 0,$$

but, in general,

$$Ext_{Ph(\mathbf{C})}^1(V, V)$$

is infinite dimensional.

(i) Consider the free particle, i.e. the dynamical system, σ given by,

$$L = 1/2\, dx^2, \quad \sigma : \delta^2 x = 0,$$

and let V be defined by letting dx act as the identity. Then we find that,

$$[\delta] = 0, Q = \frac{\partial}{\partial x}.$$

This means that $[\delta]$ does not move V in the moduli space of V. The Hamiltonian Q defines time, and

$$exp(tQ)(f(x)) = f(x + t).$$

(ii) Consider the same dynamical system, and let V be defined by letting dx act as $\frac{\partial}{\partial x}$. Then we find that,

$$[\delta] = 0, Q = (\frac{\partial}{\partial x})^2.$$

As above, $[\delta]$ does not move V in the moduli space. The Hamiltonian Q defines time, and the time evolution looks like,

$$U(t, \psi) = exp(tQ)(\psi).$$

Introducing the Fourier transformed $\hat{\psi}$, we obtain a time evolution given by,

$$U(t, \hat{\psi}) = exp(tp^2)(\hat{\psi}).$$

(iii) Consider again the harmonic oscillator, and let the representation $V := k[x^{-1}]$ be defined by letting x act as multiplication by x^{-1}, and dx act as $\frac{\partial}{\partial x}$. Then we find that,

$$[\delta] = 0, Q = (x\frac{\partial}{\partial x}).$$

As above, $[\delta]$ does not move V in the moduli space. The eigenvectors of the Hamiltonian Q are the monomials x^{-n}, $n \geq 0$, with eigenvalues $-n$, and the time evolution looks like,

$$U(t, x^{-n}) = exp(-nt)x^{-n}.$$

Notice that,

$$[x, dx] = x^2,$$

as operators on V. Notice also that V in this case is not simple. It is, however, a limit of the finite representations, $V_n := k[x^{-1}]/(x^{-1})^n$. The representation V_2 is given by the actions,

$$x = \begin{pmatrix} 0 & 0 \\ 1 & 0 \end{pmatrix}$$

$$dx = \begin{pmatrix} 0 & 0 \\ 0 & 0 \end{pmatrix}$$

where we have chosen the basis $\{1, x^{-1}\}$ in V_2. It is clearly not simple, but it sits as a point at infinity, $t_1 = t_2 = 1 + t_3 = t_4 = 0$, $t_5 = 1$, for the (almost) versal family,

$$x = \begin{pmatrix} 0 & 1 + t_3 \\ t_5 & t_4 \end{pmatrix}$$

$$dx = \begin{pmatrix} t_1 & t_2 \\ 1 + t_3 & 0 \end{pmatrix}.$$

Example 4.11. Given a dynamical system, $\mathbf{A}(\sigma)$ and a versal family for simple representations of dimension n. Let ξ be the fundamental vectorfield defined on $U(n)$. Recall Theorem (3.4.8) and Theorem (4.2.1). There is a morphism of generalized schemes,

$$U(n) \to \mathbf{Y}_n := U(n)/(\xi)$$

The quotient "spaces" $\mathbf{Y}_n$ and $\mathbf{X}_n := Simp_n(A)/(\xi)$ are orbit spaces, where each orbit is a curve. Completing, when nessecary, $U(n)$ and/or $Simp_n(A)$, we may assume these curves complete. Restricting to an *integrable part* of $U(n)$, resp. of $Simp_n(A)$, we may then hope to find natural morphisms,

$$\Gamma_n : \mathbf{Y}_n \to \mathbf{M},$$

where $\mathbf{M}$ is the moduli space of the complete algebraic curves. Moreover, Theorem (4.2.1) should produce a rank n bundle $\mathbf{V}_n$ on $\mathbf{Y}_n$, $n \geq 1$, and one might ask for conditions for the existence of universal bundles $\mathbf{U}_n$ on $\mathbf{M}$, such that $\mathbf{V}_n = \Gamma_n^* \mathbf{U}_n$.

These are questions related to vertex algebras (bundles), see e.g. [5]. There is a large literature on the subject. Seen from our point of view, the hidden agenda of the vertex algebra framework, seems to be to construct the relevant algebra $\dot{A}(\sigma)$ of observables for a given quantum (field) theoretic situation.

In our language, let $\underline{t}_0 \in Simp_n(A(\sigma))$ be a singularity for ξ. Consider the Planck's constant, $\hbar(t_0)$, and the corresponding operators, $a_i^+, a_i^- \in A(\sigma)$, together with the *vacuum state* $\omega(\underline{t}_0) \in \tilde{V}(\underline{t}_0) =: V$ (any flat section ω of $\tilde{V}$ along γ will produce a vacuum state), such that the action of $A(\sigma)$ induces an isomorphism,

$$k[a_1^+, ..., a_r^+] \simeq V,$$

a situation that we have seen realized in the case of the harmonic oscillator in dimension 1, but which is easily seen to generalize to any dimension, then there pops up a family of generalized vertex algebras. In fact, consider the restriction of the versal family

$$\tilde{\rho} : A(\sigma) \to End_{C(n)}(\tilde{V}),$$

to the integral curve γ through the point $\underline{t}_0 \in Simp_n(A(\sigma))$. It is singular at $\underline{t}_0$, so parametrized with time, τ, the completion will produce a map,

$$Y : V = \tilde{V}(\underline{t}_0) \simeq k[a_1^+, ..., a_r^+] \subset A(\sigma) \to End_k(V) \otimes_k k[[\tau]][\tau^{-1}],$$

see (3.6), which will be a kind of *generalized vertex algebra*. In particular, the localization axiom of vertex algebras imply that $\tilde{\rho}(a_i^+)$ and $\tilde{\rho}(a_j^+)$ commute, which here is obvious. Moreover we observe that the *exponentiating* formula of Y.-Z. Huang, see [5], p.18, (16),

$$Y(a, t) = R(\rho)Y(R(\rho(t)^{-1}a, \rho(t))R(\rho)^{-1},$$

for $a \in A$, and for any $\rho \in Aut(\hat{O}_{\gamma,0}) \simeq Aut(\mathbf{C}[[t]])$, follows from Theorem (4.2.3) above. We shall, hopefully, return to this in a later paper.

4.8　Clocks and Classical Dynamics

Going back to *(4.5) General Quantum Fields, Lagrangians and Actions*, we shall study the 1-dimensional case from a different perspective.

When we talk about a clock, we obviously do not talk about *the clock*. We just think of a device that can measure the changes that we choose to study, in a most objective way.

The western way of thinking about *time* is related to the old dichotomy; the past that has been, and is no more, and the future that is not yet; split by the present. We are talking about 12-18 billion years after the Big Bang, and maybe of an infinite future for our Universe. Time is maybe starting, but not necessarily ending. It has no structure, like space; it is freely flowing.

The measuring device must therefore (?) be modeled by a one-dimensional free particle, i.e. the line $k[\tau]$ with dynamical structure given by $d^2\tau = 0$.

The eastern way of thinking about time has always been cyclical, life, death, reincarnation, new death, etc. This way of viewing the world would be more comfortable with a Big Crunch turning into a new Big Bang, and so on.

The measuring device for this kind of time-notion might therefore be a one-dimensional harmonic oscillator i.e. $k[\tau]$ with dynamical structure $d^2\tau = \tau$.

Our representations of a western clock in rank 1 is easy. We have a k-algebra $k < \tau, d\tau >$ with a one-dimensional automorphism, given by, $exp(t\delta)(f)(\tau, d\tau) = f(\tau_0 + td\tau, d\tau)$. In rank 2 we may look at the situation in (4.3), and we find that the western clock has no singularities in rank 2. It never stops, in contrast to what we found for the eastern clock, see (4.3).

Let now $A := k[\tau]$, $B := k[t_1, ..., t_m]$, and let the dynamical system σ defined on A be the Eastern Clock, and the dynamical system μ on B be the free particle, so that $A(\sigma) = Ph(A)$, $d^2\tau = \tau$, $B(\mu) = Ph(B), d^2 t_i = 0$, $i = 1, ..., n$. A field, $\phi : A \rightarrow B$, should be considered as a model for the physical phenomenon that can be observed everywhere on $Simp_1(B)$, having a value that oscillate, (real or complex situation). The equation of motion for this field is,

$$d\phi(a) - \phi(da) = [Q, \phi(a)], \ a \in A(\sigma), \ Q \in B(\mu).$$

Plugging in $a = d\tau$, we obtain,

$$d^2\phi(\tau) - \phi(\tau) = [Q, \phi(d\tau)],$$

in $B(\mu)$. In the representation $Ham : B(\mu) \rightarrow B_0(\mu) = k[t_i, dt_j]$, this gives us,

$$\sum_{i,j} \frac{\partial^2 \phi}{\partial t_i \partial t_j} k_i k_j = \phi,$$

where $k_i = Ham(dt_i)$, necessarily are constants, since $d(dt_i) = 0$. This choice of vector $\mathbf{k} = (k_1, ..., k_m)$, is now arbitrary, and corresponds to the vector that pop up in physicists text, in *Fourier expansions*. We shall show that this immediately leads to the quantized Klein-Gordon equation, but first we have to make clear where we are.

4.9 Time-space and Space-times

Go back to our basic model, $Hilb^2(\mathbf{E}^3) := \mathbf{H} = \tilde{H}/Z_2$, classifying the family of pairs of points (o, x) of the Euclidean 3-space, $\mathbf{E}^3$. As above we shall first consider the structure of $\tilde{H}$, before we extend the results to $Hilb^2$. Recall from the Introduction, that given a metric g on $\tilde{H}$, there are two 3-dimensional distributions, normal to each other, one being the canonical 0-velocities, $\tilde{\Delta}$, and the other being the light velocities $\tilde{c}$. By definition, g is the time, and it is easy to see that $g_{\tilde{\Delta}}$ is the *proper time* of Einstein's relativity theory. The group of isometries of $\tilde{H}$, leaving $\tilde{\Delta}$ stable, does not contain the Lorentz boosts, $K = J^{0,1}, J^{0,2}, J^{0,3}$. This, together with the results of the section *Connections and the Generic Dynamical Structure* explains the fact that K is not concerved, and why we do not use the eigenvalues of K to label physical states, see [30],I, 2.4, p. 61.

Moreover, the tangent bundle $T(\tilde{H})$, outside of $\underline{\Delta}$ is decomposed into the sum of the basic tangent bundles $B_o, B_x, A_{o,x}$, each of rank 2. Recall also that $A_{o,x}$ is decomposed into a *unique 0-velocity*, dual to $< dt_0 >$ and a light velocity dual to $< dt_3 >$. The sub-bundle $\mathbf{S}_{o,x}$, given by the triples $(\psi, -\psi, \phi) \in B_o \oplus B_x \oplus A_{o,x}$, in which the pair $(\psi, -\psi)$ corresponds to a light-velocity, is at each point of $\underline{H}$ a 4-dimensional tangent sub-space, with a unique 0-velocity.

Consider the symmetry in $\tilde{\underline{H}}$, induced by the generator $\tau \in Z_2$. The tangent space of $\tilde{\underline{H}}$ at the point $\underline{t} \notin \underline{\Delta}$ is represented by the vector space of pairs $\{(\xi_o, \xi_x)\}$ where ξ_o is a tangent vector in the Euclidean 3-space at the point o, and ξ_x is a tangent vector at the point x. Any such pair may be written as,

$$(\xi_o, \xi_x) = (1/2(\xi_o + \xi_x), 1/2(\xi_o + \xi_x)) + (1/2(\xi_o - \xi_x), 1/2(-\xi_o + \xi_x)),$$

where the first vector is in $\tilde{\Delta}$, and the second in $\tilde{c}$. Clearly τ leaves $\tilde{\Delta}$ fixed and is the multiplication by -1 on $\tilde{c}$. τ also inverts *chirality*, and *spin*, since the orientation of B_o, defined by (o, x) is the the inverse of the orientation of B_x defined by (x, o).

Classically one defines the symmetry operators in Minkowski space, the *parity operator P*, by multiplying the space-coordinates with -1, the *time inversion operator T*, by multiplying the time coordinate with -1, and the *charge conjugation operator C*, by multiplying the *spin* σ by -1. Identifying $\tilde{c}$ with the past light-cone in Minkowski space, we see that τ corresponds to the transform $(x_o, x_1, x_2, x_3) \to (-x_o, -x_1, -x_2, -x_3)$, i.e. τ corresponds to PT, so $\tau^2 = \tau PT = id$. This suggests that τ is the charge conjugation operator, but as we have seen it inverts both spin and the momenta in light-direction, so it is (slightly?) different from the classically defined C, see [30], I, (3.3), p.131.

Choosing a line $l \subset \mathbf{E}^3$, the subscheme $\underline{H}(l) \subset \underline{H}$, has a much simpler structure than $\underline{H}$. The sub-bundle $\mathbf{S}_{o,x}$, restricted to $\underline{H}(l)$ can be integrated in $\underline{H}$, and we obtain a 4-dimensional subspace $\underline{S}(l) \subset \underline{H}$, in which we choose coordinates t_0, dt_1, dt_2, t_3, where dt_0 and dt_3 are as above, and dt_1, dt_2 are dual coordinates for the *transverse* bundle, B_o, (isomorphic to the inversely oriented bundle B_x), normal to l in $\mathbf{E}^3$.

This subspace $S(l)$ of $\underline{H}$ may be identified with a natural moduli-subspace $M(l) \subset \tilde{\underline{H}}$. In fact, let $M(l) = \{(o, x) \in \tilde{\underline{H}} | 1/2(o + x) \in l\}$.

Since $o = 1/2(o + x) + 1/2(o - x), x = 1/2(o + x) - 1/2(o - x)$, it is of dimension 4, and contains $H(l)$. At every point $(o, x) \in \underline{H}(l) \subset M(l)$ the tangent space is easily identified with $\mathbf{S}_{o,x}$, therefore identifying $S(l)$ and $M(l)$, as spaces. However, the structure of $M(l)$ is much richer than that of $S(l) \simeq \underline{S}(l) \times \mathbf{A}^2$. In particular, the actions of the gauge group are different, see (Example 4.14). See also the section *Cosmology, Big Bang and all that*, for another characterization of $S(l)$.

Moreover, the usual Minkowski space, is recovered as the restriction, $U(l)$, of the universal family $\underline{U}$, to $M(l)$. The metric is deduced from the *energy-function* of $M(l)$.

Now go back to (4.8) and substitute $S(l)$ for B. We have two solutions for ϕ, respectively, $\phi = exp(\mathbf{kt}), \phi^\dagger = exp(-\mathbf{kt})$, in the real case, and with $\delta = \imath d$ substituting for d, $\phi = exp(\imath \mathbf{kt}), \phi^\dagger = exp(-\imath \mathbf{kt})$, in the complex case. Now, in $S(l)(\mu)$ the elements ϕ and $\phi^\dagger$ commute, but they do not commute with $d\phi$ and $d\phi^\dagger$. However, if we now look at the representation,

$$Wey : S(l)(\mu) \to Diff_k(S(l)),$$

where we recall that,

$$Diff_k(S(l)), = k < t_i, dt_j > /([t_i, t_j], [dt_i, dt_j], [t_i, dt_j]_{i \neq j}, [t_i, dt_i] - 1),$$

we find, $[dt_i, \phi] = k_i \phi$, $[dt_i, \phi^\dagger] = -k_i \phi^\dagger$, and computing a little,

$$d\phi = \phi \sum_i k_i dt_i + 1/2\phi \sum_i k_i^2$$

$$d\phi^\dagger = -\phi^\dagger \sum_i k_i dt_i + 1/2\phi^\dagger \sum_i k_i^2$$

$$[d\phi, \phi^\dagger] = -\sum_i k_i^2 = -|\mathbf{k}|^2, \quad [d\phi^\dagger, \phi] = -\sum_i k_i^2 = -|\mathbf{k}|^2.$$

In fact, let $K := \mathbf{kt} = \sum_i k_i t_i$, and put $L := \mathbf{k}^2 = \sum_i k_i^2$, then, $dK = \sum_i k_i dt_i$, and we have,

$$d(K^p) = pK^{p-1}dK + (1/2)(p - 1)pK^{p-2}L$$

$$\phi = exp(K), \quad \phi^\dagger = exp(-K),$$

$$d\phi = \phi dK + (1/2)L\phi, \quad d\phi^\dagger = -\phi^\dagger dK + (1/2)L\phi^\dagger.$$

Put $Q := \sum_i k_i dt_i$, and see that in $\underline{H}$, and so also in $\underline{S}(l)$, Q is the energy operator, along $l := \mathbf{k}$. A simple computation shows that,

$$\phi^\dagger d\phi = Q + 1/2L$$

$$\phi d\phi^\dagger = -Q + 1/2L,$$

and since $\phi\phi^\dagger = 1$, also,

$$d\phi^\dagger\phi = -Q - 1/2L$$
$$d\phi\phi^\dagger = Q - 1/2L.$$

This is, in essence, the *quantized Klein-Gordon*, as one finds it in new textbooks on QFT, see e.g. [23]. Notice also that, from Example (4.3) we know that the energy of this *particle* is given in terms of the number operator, here we have, $Q = -1/2(\phi d\phi^\dagger + d\phi^\dagger\phi) = 1/2(\phi^\dagger d\phi + d\phi\phi^\dagger)$.

Example 4.12. Newton's and Kepler's Laws

Let us study the geometry of **H**. Recall that $\tilde{\underline{H}} \to \underline{H}$, is the (real) blow up of the diagonal $\underline{\Delta} \subset \underline{H}$, where $\underline{H}$ is the space of pairs of points in $\mathbf{E}^3$. Clearly any point $\underline{t} \in \underline{H}$ outside the diagonal, determines a vector $\xi(o, x)$ and an oriented line $l(o, x) \subset \mathbf{E}^3$, on which both the observer o and the observed x sits. This line also determines a subscheme $\underline{H}(l) \subset \underline{H}$, see above and [20], and in $\underline{H}(l)$ there is unique *light velocity curve* $\underline{l}(\underline{t})$, through $\underline{t}$, an integral curve of the distribution $\tilde{c}$, and this curve cuts the diagonal $\underline{\Delta}$ in a unique point $c(o, x)$, the *center of gravity of the observer and the observed*, and thus defines a unique point $\xi(\underline{t})$, of the blow-up of the diagonal, in the fiber of $\tilde{\underline{H}} \to \underline{H}$, above $c(o, x)$. Any tangent $\eta := (\eta_1, \eta_2), \eta_2 = -\eta_1$, of $\underline{H}$ in $\tilde{c}$, at $\underline{t} = (o, x)$, normal to $\underline{l}(\underline{t})$, corresponds to a light velocity, to a *spin vector*, $\eta_1 \times \xi(o, x)$, in $\mathbf{E}^3$, with *spin axis*, the corresponding oriented line. The length of the spin vector is called the *spin* of η. We have shown in [20] that there exists a metric on $\underline{H}$ which, restricted to every 3-space $\underline{c}(\underline{t}) - \{c(o, x)\}$, has the form,

$$ds^2 = dt_3^2 + (1 + r^{-2})dt_1^2 + (1 + r^{-2})dt_2^2 - r^{-4}(t_1 dt_1 + t_2 dt_2)^2,$$

where we have chosen the coordinates such that dt_3 corresponds to the oriented line $l(o, x)$, and dt_i, $i = 1, 2$, correspond to the spin-momenta, assuming $t_3 \neq 0$. The nice property of this metric is the following. Consider a spin momentum $\eta := (\eta_1, -\eta_1)$ and its corresponding spin vector $\eta \times \xi(o, x) := (\eta_1 \times \xi(o, x), -\eta_1 \times \xi(o, x))$ along the line $\underline{l}(\underline{t})$. Clearly the length of this vector, when $r = t_3$ is large, is just the classical spin. When r tends to zero, η defines a tangent vector of the exceptional fiber of the blow up at $c(o, x)$, i.e. of the projective 2-space, and of the covering 2-sphere. And we see that the length of $\eta \times \xi(o, x)$ tends to the length of this tangent vector in the Fubini-Studi metric of $\mathbf{P}^2$.

To see what this may lead us to, we need a convenient parametrization of $\tilde{\underline{H}}$. Consider, as above, for each $\underline{t} \in \tilde{\underline{H}}$ the length ρ, in $\mathbf{E}^3$, the Euclidean space, of the vector (o, x). Given a point $\underline{\lambda} \in \underline{\Delta}$, and a point $\xi \in E(\underline{\lambda}) = \pi^{-1}(\underline{\lambda})$, the fiber of,

$$\pi : \underline{\tilde{H}} \to \underline{H},$$

at the point $\underline{\lambda}$, for $o = x$. Since $E(\underline{\lambda})$ is isomorphic to S^2, parametrized by $\underline{\phi}$, any element of $\underline{\tilde{H}}$ is now uniquely determined in terms of the triple $\underline{t} = (\underline{\lambda}, \underline{\phi}, \rho)$, such that $c(\underline{t}) = c(o, x) = \underline{\lambda}$, and such that ξ is defined by the line $\underline{ox}$. Here $\rho \geq 0$, see also the section *Cosmology, Big Bang and all that.* Notice also that, at the exceptional fiber, i.e. for $\rho = 0$, the momentum corresponding to $d\rho$ is not defined.

Consider any metric on $\underline{\tilde{H}}$, of the form,

$$g = h_1(\underline{\lambda}, \underline{\phi}, \rho)d\rho^2 + h_2(\underline{\lambda}, \underline{\phi}, \rho)d\underline{\phi}^2 + h_3(\underline{\lambda}, \underline{\phi}, \rho)d\underline{\lambda}^2,$$

where $d\underline{\phi}^2$ is the natural metric in $S^2 = E(\underline{\lambda})$. It is reasonable to believe that the geometry of $(\underline{\tilde{H}}, g)$, might explain the notions like *energy*, mass, *charge*, etc.. In fact, we tentatively propose that the source of mass and charge etc. is located in the *black holes* $E(\underline{\lambda})$. This would imply that mass, charge, etc. are properties of the 5-dimensional superstructure of our usual 3-dimensional Euclidean space, essentially given by a *density*, $h(\underline{\lambda}, \underline{\phi}, \theta)$. This might bring to mind Kaluza-Klein-theory. However, it seems to me that there are important differences, making comparison very difficult.

Here we shall just consider the following simple case, where

$$h_1 = (\frac{\rho - h}{\rho})^2, \quad h_2 = (\rho - h)^2, \quad h_3 = 1,$$

where h is a positive real number. This metric is everywhere defined, since for $\rho = 0$, there are no tanget vectors in $d\rho$ direction. It clearly reduces to the Euclidean metric far away from $\underline{\Delta}$, and it is singular on the *horizon* of the black hole, given by $\rho = h$, which in $\underline{H}$ is simply a sphere in the *light-space*, of radius h. Moreover it is clear that h is also the *radius of the exceptional fibre*, since the length of the circumference of $\rho = 0$, is $2\pi h$. Clearly, the exceptional fiber, the black hole itself, is not visible, and does not bound anything. However, the horizon bounds a piece of space. Moreover, if we reduce the horizon to a point in $\underline{H}$, then the circumference, or area of the exceptional fiber, as measured using the above metric, reduces to zero, and the metric becomes the usual Euclidean metric.

We shall reduce to a plane in the light directions, i.e. we shall just assume that $S^2 = E(\underline{\lambda})$, is reduced to a circle, with coordinate ϕ. This is actually no restriction made, as is easily seen. Consider the Lagrangian

$L = g$, see (Example 4.1), we find the Euler-Lagrange equations,

$$0 = \delta(\frac{\partial L}{\partial d\rho}) - \frac{\partial L}{\partial \rho} = 2(\frac{\rho - h}{\rho})^2 d^2\rho + 2(\frac{\rho - h}{\rho})(\frac{h}{\rho^2})d\rho^2 - 2(\rho - h)d\phi^2$$

$$0 = \delta(\frac{\partial L}{\partial d\phi}) - \frac{\partial L}{\partial \phi} = 2(\rho - h)^2 d^2\phi + 4(\rho - h)d\rho d\phi$$

$$0 = \delta(\frac{\partial L}{\partial d\lambda}) - \frac{\partial L}{\partial \lambda} = 2d^2\lambda$$

where $\lambda = \underline{\lambda}$, as above.

We solve these equations, and find,

$$d^2\rho = -(\frac{h}{\rho(\rho - h)})d\rho^2 + (\frac{\rho^2}{(\rho - h)})d\phi^2$$

$$d^2\phi = -2/(\rho - h)d\rho d\phi$$

$$d^2\lambda = 0.$$

This, of course, is the same solutions as what we would have found by solving the Lagrange equation.

According to Example (4.1) the corresponding equations for the geodesics in $\tilde{H}$ are,

$$\frac{d^2\rho}{dt^2} = -(\frac{h}{\rho(\rho - h)})(\frac{d\rho}{dt})^2 + (\frac{\rho^2}{(\rho - h)})(\frac{d\phi}{dt})^2,$$

$$\frac{d^2\phi}{dt^2} = -2/(\rho - h)\frac{d\rho}{dt}\frac{d\phi}{dt}$$

$$\frac{d^2\lambda}{dt^2} = 0,$$

where t is time. But time is, by definition, the distance function in $\tilde{\underline{H}}$, so we must have,

$$(\frac{\rho - h}{\rho})^2(\frac{d\rho}{dt})^2 + (\rho - h)^2(\frac{d\phi}{dt})^2 + (\frac{d\lambda}{dt})^2 = 1,$$

from which we find,

$$(\frac{d\rho}{dt})^2 = \rho^2(\rho - h)^{-2}(1 - (\frac{d\lambda}{dt})^2) - \rho^2(\frac{d\phi}{dt})^2.$$

From the third equation, we find that $\frac{d\lambda_j}{dt}$, $j = 1, 2, 3$, are constants, and $|\frac{d\lambda}{dt}|$ is the *rest-mass* of the system. Put $K^2 = (1 - |\frac{d\lambda}{dt}|^2)$, then K is the *kinetic energy of the system*. The definition of time therefore give us,

$$\rho^{-2}(\frac{d\rho}{dt})^2 = (\rho - h)^{-2}K^2 - (\frac{d\phi}{dt})^2.$$

Put this into the first equation above, and obtain,

$$\frac{d^2\rho}{dt^2} = -hK^2\left(\frac{\rho}{\rho - h}\right)\frac{1}{(\rho - h)^2} + \left(\frac{\rho + h}{\rho - h}\right)\rho\left(\frac{d\phi}{dt}\right)^2.$$

Assume now $r := \rho - h \approx \rho$, we find,

$$\frac{d^2r}{dt^2} = -\frac{hK^2}{r^2} + r\left(\frac{d\phi}{dt}\right)^2,$$

i.e. Keplers first law. The constant h, i.e. the radius of the exceptional fiber, is thus also related to mass. In fact, this suggests that *mass*, is a property of the space $\tilde{\underline{H}}$. In this case it is a function of the surface of the exceptional fiber, i.e. the *black hole*, associated with the point $\underline{\lambda}$ in the ordinary 3-space $\underline{\Delta}$.

In the same way, the second equation above gives us Keplers second law,

$$r\left(\frac{d^2\phi}{dt^2}\right) + 2\left(\frac{dr}{dt}\right)\left(\frac{d\phi}{dt}\right) = 0.$$

Notice that with the chosen metric, time, in light velocity direction, is *standing still* on the *horizon* $\rho = h$, of the *black hole* at $\underline{\lambda} \in \underline{\Delta}$. Therefore no light can escape from the black hole. In fact, no geodesics can pass through $\rho = h$. Notice also that, for a photon with light velocity, we have $K = 1$, so we may measure h, by measuring the trajectories of photons in the neighborhood of the *black hole*.

Finally, see that if the distance between the two interacting points is close to constant, i.e. if we have a circular movement, the left side of the time-equation becomes zero, and we therefore have the following equation,

$$\rho d\phi = K dt + h d\phi,$$

which may be related to the perihelion praecicion.

Let us now go back, and consider, in this case, the generic dynamical structure (σ), of the subsection *Connections, and the Generic Dynamical Structure*, related to the above metric. Put $\rho = t_1, \phi = t_2, \lambda = t_3$, then the Euler-Lagrange equations above give us immediately the following formulas,

$$\Gamma^1_{1,1} = h/\rho(\rho - h), \ \Gamma^1_{2,2} = -\rho^2/(\rho - h)$$
$$\Gamma^2_{1,2} = 1/(\rho - h), \ \Gamma^2_{2,1} = 1/(\rho - h)$$
$$\Gamma^3_{i,j} = 0$$

All other components vanish. From this we find the following formulas,

$$\nabla_1 = \begin{pmatrix} h/\rho(\rho - h) & 0 & 0 \\ 0 & 1/(\rho - h) & 0 \\ 0 & 0 & 0 \end{pmatrix}$$

$$\nabla_2 = \begin{pmatrix} 0 & 1/(\rho - h) & 0 \\ -\rho^2/(\rho - h) & 0 & 0 \\ 0 & 0 & 0 \end{pmatrix}$$

$$[\nabla_1, \nabla_2] = \begin{pmatrix} 0 & -1/\rho(\rho - h) & 0 \\ -\rho/(\rho - h) & 0 & 0 \\ 0 & 0 & 0 \end{pmatrix}$$

$$\frac{\partial}{\partial \rho} \nabla_2 = (\rho - h)^{-2} \begin{pmatrix} 0 & -1 & 0 \\ -\rho(\rho - 2h) & 0 & 0 \\ 0 & 0 & 0 \end{pmatrix}$$

$$D_\rho := \nabla_{\delta_1} = \frac{\partial}{\partial \rho} + \nabla_1,$$

$$D_\phi := \nabla_{\delta_2} = \frac{\partial}{\partial \phi} + \nabla_2,$$

$$D_\lambda := \nabla_{\delta_3} = \frac{\partial}{\partial \lambda} + \nabla_3$$

$$Q = \sum_{i=1}^{3} 1/h_i \nabla_{\delta_i}^2$$

$$\rho(\delta^2(t_i)) = [Q, \rho(dt_i)] = 1/h_i [Q, \nabla_{\delta_i}].$$

Here the h_i is the function defined above, i.e. $g_{i,i}$ in our metric.

Recall our *General Force Law* for the proposed metric on $\tilde{H}$, and the Levi-Civita connection,

$$\rho_E(d^2 t_i) + \sum_{p,q} \Gamma_{p,q}^i \nabla_{\xi_p} \nabla_{\xi_q}$$

$$= 1/2 \sum_p F_{p,i} \nabla_{\delta_p} + 1/2 \sum_p \nabla_{\delta_p} F_{p,i}$$

$$+ 1/2 \sum_{l,q} \delta_q (\Gamma_l^{i,q} - \Gamma_l^{q,i}) \nabla_{\xi_l} + [\nabla_{\xi_i}, \rho_E(T)].$$

A dull, and lengthy computation gives the formulas,

$$\xi_\rho = 2\rho^2(\rho - h)^{-2}\frac{\partial}{\partial\rho}, \ \xi_\phi = 2(\rho - h)^{-2}\frac{\partial}{\partial\phi}, \ \xi_\lambda = 2\frac{\partial}{\partial\lambda},$$

$$[\xi_\rho, \xi_\phi] = -8\rho^2(\rho - h)^{-5}\frac{\partial}{\partial\phi}$$

$$F(\xi_\rho, \xi_\phi) = 4\rho^2(\rho - h)^{-4}([\nabla_\rho, \nabla_\phi] + \frac{\partial}{\partial\rho}\nabla_\phi).$$

Putting these formulas together gives the expressions,

$$\rho_\Theta(d^2\rho) + h\rho^{-1}(\rho - h)^{-1}\nabla^2_{\xi_\rho} - \rho^2(\rho - h)^{-1}\nabla^2_{\xi_\phi}$$

$$= F(\xi_\phi, \xi_\rho)(\frac{\partial}{\partial\phi} + \nabla_\phi) + 2\rho^2(\rho - h)^{-5}(\frac{\partial}{\partial\phi} + \nabla_\phi) + [\nabla_{\xi_\rho}, \rho_\Theta(T)],$$

and,

$$\rho_\Theta(d^2\phi) + (\rho - h)^{-1}\nabla_{\xi_\phi}\nabla_{\xi_\rho} - \nabla_{\xi_\rho}\nabla_{\xi_\phi}$$

$$= 1/2F(\xi_\rho, \xi_\phi)(\frac{\partial}{\partial\rho} + \nabla_\rho) + 1/2(\frac{\partial}{\partial\rho} + \nabla_\rho)F(\xi_\rho, \xi_\phi)$$

$$- (4\rho(\rho - h)^{-5} + 6\rho^2(\rho - h)^{-6})(\frac{\partial}{\partial\phi} + \nabla_\phi) + [\nabla_{\xi_\phi}, \rho_\Theta(T)].$$

Example 4.13. Classical Electro-Magnetism There are at least two possible models of an electromagnetic field.

First, given a potential,

$$\phi = \sum_{j=0}^{3} \phi_j d_1 t_j \in PhS(l),$$

considered as a field $\phi : k[\tau] \to Ph(S(l))$ and, say the trivial metric g on $\underline{S}(l)$. The interpretation is a plane wave, corresponding to a linear form on the tangent space of each point of $\underline{S}(l)$. Notice that we have a canonical derivation $d_1 : S(l) \to Ph(S(l))$, and that we have the following relations,

$$[d_1 t_i, t_j] + [t_i, d_1 t_j] = 0.$$

Let $A := k[\tau]$ be the Western clock, with Dirac derivation d with $d^2\tau = 0$. It is easy to check that the following relations actually define a dynamic system on $B = Ph(S(l))$, with Dirac derivation, d:

$$[t_i, dt_j] = 0, i, j \geq 0,$$
$$dt_i d_1 t_j = -dt_j d_1 t_i, i \neq j,$$
$$dt_i d_1 t_i = dt_j d_1 t_j, i, j \geq 0,$$
$$d^2 t_i = dd_1 t_i = 0, \ i = 0, 1, 2, 3.$$

If we put, for $i, j, k = 1, 2, 3$, with $sgn(i, j, i \times j) = 1$,

$$E_i := \frac{\partial \phi_i}{\partial t_0} - \frac{\partial \phi_0}{\partial t_i}, \ B_k := \frac{\partial \phi_j}{\partial t_i} - \frac{\partial \phi_i}{\partial t_j}, k = i \times j,$$

we find, modulo these relations,

$$d\phi = E_1 dt_0 d_1 t_1 + E_2 dt_0 d_1 t_2 + E_3 dt_0 d_1 t_3 + B_3 dt_1 d_1 t_2 + B_2 dt_3 d_1 t_1 + B_1 dt_2 d_1 t_3$$

$$+ \nabla.\phi dt_0 d_1 t_0 + \sum_{i=0}^{3} \phi_i dd_1 t_i.$$

Computing we find,

$$\delta^2(\phi) = \left(\frac{\partial(\nabla.\phi)}{\partial t_0} + \frac{\partial E_1}{\partial t_1} + \frac{\partial E_2}{\partial t_2} + \frac{\partial E_3}{\partial t_3} \right) dt_0 dt_0 d_1 t_0$$

$$+ \left(\frac{\partial(\nabla.\phi)}{\partial t_1} + \frac{\partial E_1}{\partial t_0} + \left(\frac{\partial B_2}{\partial t_3} - \frac{\partial B_3}{\partial t_2} \right) \right) dt_1 dt_0 d_1 t_0$$

$$+ \left(\frac{\partial(\nabla.\phi)}{\partial t_2} + \frac{\partial E_2}{\partial t_0} + \left(\frac{\partial B_3}{\partial t_1} - \frac{\partial B_1}{\partial t_3} \right) \right) dt_2 dt_0 d_1 t_0$$

$$+ \left(\left(\frac{\partial \nabla.\phi)}{\partial t_3} + \frac{\partial E_3}{\partial t_0} + \left(\frac{\partial B_1}{\partial t_2} - \frac{\partial B_2}{\partial t_1} \right) \right) dt_3 dt_0 d_1 t_0$$

$$+ \left(\left(\frac{\partial E_1}{\partial t_2} - \frac{\partial E_2}{\partial t_1} \right) + \frac{\partial B_3}{\partial t_0} \right) dt_0 dt_1 d_1 t_2$$

$$+ \left(\left(\frac{\partial E_1}{\partial t_3} - \frac{\partial E_3}{\partial t_1} \right) - \frac{\partial B_2}{\partial t_0} \right) dt_0 dt_1 d_1 t_3$$

$$+ \left(\left(\frac{\partial E_2}{\partial t_3} - \frac{\partial E_3}{\partial t_2} \right) + \frac{\partial B_1}{\partial t_0} \right) dt_0 dt_2 d_1 t_3$$

$$+ \left(\frac{\partial B_3}{\partial t_3} + \frac{\partial B_2}{\partial t_2} + \frac{\partial B_1}{\partial t_1} \right) dt_1 dt_2 d_1 t_3$$

$$+ D.$$

Here,

$$D = \left(\frac{\partial \phi_0}{\partial t_1} - \frac{\partial \phi_1}{\partial t_0} \right) dt_1 d^2 t_0 + \left(\frac{\partial \phi_0}{\partial t_2} - \frac{\partial \phi_2}{\partial t_0} \right) dt_2 d^2 t_0$$

$$+ \left(\frac{\partial \phi_0}{\partial t_3} - \frac{\partial \phi_3}{\partial t_0} \right) dt_3 d^2 t_0 + \left(\frac{\partial \phi_1}{\partial t_2} - \frac{\partial \phi_2}{\partial t_1} \right) dt_2 d^2 t_1$$

$$+ \left(\frac{\partial \phi_1}{\partial t_3} - \frac{\partial \phi_3}{\partial t_1} \right) dt_3 d^2 t_1 + \left(\frac{\partial \phi_2}{\partial t_3} - \frac{\partial \phi_3}{\partial t_2} \right) dt_3 d^2 t_2$$

$$+ \sum_{i=0}^{3} \frac{\partial \phi_i}{\partial t_i} dt_i d^2 t_i,$$

vanish. The equation of motion is, $d^2\phi = 0$, which implies the following equations,

$$\frac{\partial(\nabla.\phi)}{\partial t_0} + \nabla_s.E = 0$$

$$\nabla_s(\nabla.\phi) + \frac{\partial E}{\partial t_0} + \nabla_s \times B = 0$$

$$\nabla_s \times E + \frac{\partial B}{\partial t_0} = 0$$

$$\nabla_s.B = 0,$$

where ∇_s is the space-gradient.

These are Maxwell's equations, with electrical charge equal to $\rho := \frac{\partial(\nabla.\phi)}{\partial t_0}$, and electrical current equal to $\mathbf{j} := \nabla_s.(\nabla.\phi)$. The last two equations are Bianchi's equations and are trivial, given that we start with a potential.

In our language, we see that charge becomes a rest-mass, and rest-mass a kinetic energy.

Notice also that in space-time coordinates our potential satisfies, $\nabla^2\phi_i = 0$, which explains the extra conditions necessary in the classical case. A classical *free* particle, clocked by a Western clock, is now, according to Example (4.13) a field

$$\kappa : S(l) \to k[\tau]$$

such that,

$$\delta S = 0, \ S := \int L, d\tau$$

where, putting $\dot\kappa_j := \frac{\partial \kappa_j}{\partial \tau}$,

$$L := Ph(\kappa)(\phi) = \sum_{j=0}^{3} \phi_j(\kappa_0, \kappa_1, \kappa_2, \kappa_3)\dot\kappa_j.$$

Classically, where the representations one considers are L^2-spaces of functions, this is interpreted as,

$$\delta \int Ph(\kappa)(\phi)d\tau = 0.$$

The Euler-Lagrange equations applies and one gets the system of equations,

$$\frac{\partial \phi_j}{\partial \tau} - \sum_i \frac{\partial \phi_i}{\partial t_j}\dot\kappa_i = 0, \forall j,$$

which reduces to,

$$\sum_{i=0}^{3} \frac{\partial \phi_j}{\partial t_i} \dot{\kappa}_i - \sum_{i=0}^{3} \frac{\partial \phi_i}{\partial t_j} \dot{\kappa}_i = 0, \forall j.$$

Put, as above,

$$E_i := \frac{\partial \phi_i}{\partial t_0} - \frac{\partial \phi_0}{\partial t_i}, \ B_k := \frac{\partial \phi_j}{\partial t_i} - \frac{\partial \phi_i}{\partial t_j}, k = i \times j,$$

and define the 3-vectors,

$$\psi := (\kappa_1, \kappa_2, \kappa_3), \ \dot{\psi} := (\dot{\kappa}_1, \dot{\kappa}_2, \dot{\kappa}_3)$$
$$E := (E_1, E_2, E_3), \ B := (B_1, B_2, B_3).$$

Then the equations above simply says the following,

$$B \times \dot{\psi} + E.\dot{\kappa}_0 = 0, \ E.\dot{\psi} = 0.$$

Now let us denote by γ the curve defined by $\kappa = (\kappa_0, \kappa_1, \kappa_2, \kappa_3)$. Then in our space $\underline{S}(l) \subset \underline{H}$ the squared energy, m^2, of the particle, measured by the clock we are using, is $\sum_{i=0}^{3} \dot{\kappa}_i^2$. Recalling that,

$$\dot{\gamma} = m.\mathbf{v},$$

where the unit vector $\mathbf{v}$ is the velocity, and $\dot{\psi} = m\mathbf{v}_{space}$, $\dot{\kappa}_0 = m.\mathbf{v}_0 :=$ *restmass*. We find the following equations,

$$B \times \mathbf{v}_{space} + E.\mathbf{v}_0 = 0.$$

If we, together with the electromagnetic potential, also take into consideration gravitation, i.e. time, say just the trivial metric, $g := \sum_{i=0}^{3} dt_i^2$, i.e. if we consider the Lagrangian,

$$L := \phi + g,$$

then, the Euler-Lagrange equations look like,

$$\frac{\partial \dot{\kappa}_j}{\partial \tau} + \frac{\partial \phi_j}{\partial \tau} - \sum_i \frac{\partial \phi_i}{\partial t_j} \dot{\kappa}_i = 0, \forall j.$$

Referring to Example (4.12), we find for the real time, t, with respect to the clock-time, τ,

$$(\frac{\partial t}{\partial \tau})^2 = \sum_{i=0}^{3} \dot{\kappa}_i^2, \ \frac{\partial \mathbf{v}}{\partial \tau} = \dot{\mathbf{v}} = \frac{\partial \mathbf{v}}{\partial t} \frac{\partial t}{\partial \tau} = \mathbf{a}.m,$$

where **a** is the relativistic acceleration, and the equations above become,

$$ma = m \begin{pmatrix} 0 & E_1 & E_2 & E_3 \\ -E_1 & 0 & B_3 & -B_2 \\ -E_2 & -B_3 & 0 & B_1 \\ -E_3 & B_2 & -B_1 & 0 \end{pmatrix} \begin{pmatrix} \mathbf{v}_0 \\ \mathbf{v}_1 \\ \mathbf{v}_2 \\ \mathbf{v}_3 \end{pmatrix}.$$

This is, basically, what one finds in any textbook in physics, see again [23]. The mass, or relativistic energy of the *test particle* here is m, and the *charge density* becomes a kind of relativistic energy density in the (unique) 0-velocity direction dt_0, both measured with our clock. Notice that this kind of rest-energy is the only energy or mass that we seem to be able to observe via electromagnetic interaction. The *missing* two other 0-velocity directions might hide black energy/mass? Notice also that **v** and therefore **a** are just dependent upon the structure of $\underline{H}$ and the curve γ, not on the clock.

With this done, one may consider the Weyl representation,

$$Ph(S(l)) \to Diff(S(l)),$$

which simply means to introduce the new relations, $[t_i, d_1 t_j] = 0$, $i \neq j$, $[t_i, d_1 t_i] = 1$. Then, going about as above, in the case of the Klein-Gordon equation, one obtains the classical quantization of Electro Magnetism, see also the next Example.

For later use, see that the last formula is a kind of Schrödinger equation,

$$\frac{\partial}{\partial t}(\psi) = Q(\psi),$$

where $\Psi \in \Theta_{\underline{S}(l)}$, and $Q \in End_{S(l)}(\Theta_{S(l)})$.

Let us now go back to the section, *Connections and the Generic Dynamical Structure*, and do all this via the interpretation of an electromagnetic field, as given, for a trivial metric, by the connection ∇, of the tangent bundle $\Theta_{\underline{S}(l)}$,

$$\nabla_{\delta_i} = \frac{\partial}{\partial t_i} + A_i.$$

Here $A_i \in End_{S(l)}(\Theta_{S(l)})$, in the above classical situation, are just functions, acting as a diagonal matrix. Notice, however, that this connection now has *torsion*.

As we have seen, there is a canonical associated representation, ρ_∇, for which the curvature is given by,

$$F_{i,j} := \rho_\nabla([dt_i, dt_j]) = \frac{\partial A_i}{\partial t_j} - \frac{\partial A_j}{\partial t_i}.$$

Since the metric is flat, the generic dynamical system, (σ), will give us,

$$d^2 t_i = -1/2 \sum_{j=1}^{r} (dt_j [dt_i, dt_j] + [dt_i, dt_j] dt_j),$$

so we have, the force law,

$$d^2 t_i = -\sum_{j=1}^{r} F_{i,j} dt_j - q_i,$$

where the vector $\underline{q}$ is the charge-current density. In this case we have,

$$q_i = 1/2 \sum_{j=1}^{r} \frac{\partial F_{i,j}}{\partial t_j}.$$

The Maxwell equations, the 2 first non-trivial ones, are then equivalent to,

$$\underline{q} = \nabla(\nabla A) - \nabla^2(A).$$

We see here that the charge occur also in the equation of motion for fields. It disappeared in the essentially commutative QF-version, presented above. As in the case of general relativity, where the problem was to explain the notion of mass, as a property of the geometry of the time-space, the problem here is to explain how the notion of charge is related to the the geometry of the same time-space. This will be treated in a forthcoming paper by Olav Gravir Imenes, see also [8].

Now, consider any Lie group G, acting on a k-algebra A. The action induces a homomorphism of Lie algebras,

$$\eta : \mathbf{g} \to Der_k(A).$$

For a fixed integer n, there is a versal family,

$$\tilde{\rho} : A \to End_{C(n)}(\tilde{V}).$$

Any element $\chi \in \mathbf{g}$, considered as a derivation of A, acts, according to Theorem (4.2), like,

$$\tilde{\rho}(\chi(a)) = [\chi](\tilde{\rho}(a)) + [\nabla_\chi, \tilde{\rho}(a)],$$

for $a \in A$, and for some Hamiltonian ∇_χ. This looks like a connection,

$$\nabla : \mathbf{g} \to End_k(\tilde{V}).$$

Clearly, the condition $[\chi] = 0$ for all $\chi \in \mathbf{g}$, implies that the action η induces an $A - \mathbf{g}$-module structure on $\tilde{V}$,

$$\nabla : \mathbf{g} \to End_{C(n)}(\tilde{V}),$$

as in [18], p.563. This means that ∇ is a Lie-algebra homomorphism, and that for all $a \in A$, $\chi \in \mathfrak{g}$, $\psi \in \tilde{V}$, we have

$$\nabla_\chi(\tilde{\rho}(a)(\psi)) = \tilde{\rho}(a)\nabla_\chi(\psi) + \tilde{\rho}(\chi(a))(\psi).$$

In particular the curvature vanishes, i.e.

$$R(\chi, \eta) := [\nabla_\chi, \nabla_\eta] - \nabla_{[\chi,\eta]} = 0.$$

Moreover, we find that the sub-scheme,

$$\underline{C}(n; \mathfrak{g}) := \{c \in Simp_1(C(n)) | \forall \chi \in \mathfrak{g}, [\chi](c) = 0\},$$

is a sub-scheme of $Simp(A : \mathfrak{g})$, the non-commutative invariant space defined by the $\mathfrak{g}$-action.

In fact, what we do by imposing the conditions $[\chi] = 0$, for all $\chi \in \mathfrak{g}$, is to construct a *slice* in $Simp_1 C(n)$ cutting all orbits of $[\mathfrak{g}]$ at a single point. The result is obviously an *orbit space*, for which there exist an essentially unique $A - \mathfrak{g}$-module structure on the restriction of $\tilde{V}$ to $\underline{C}(n; \mathfrak{g})$. In particular we find a Lie-algebra homomorphism,

$$\nabla : \mathfrak{g} \to End_{\underline{C}(n;\mathfrak{g})}(\tilde{V}).$$

Conversely, if we start with some Lie-algebra homomorphism,

$$\nabla : \mathfrak{g} \to End_C(\tilde{V}),$$

and a representation,

$$\rho : A \to End_C(\tilde{V}),$$

we might hope to construct an action of $\mathfrak{g}$ on A, such that for $\xi \in \mathfrak{g}$, $a \in A$, $\rho(\xi(a)) = [\nabla_\xi, \rho(a)]$. This is, however, rarely the case.

Example 4.14. (Gauge Groups, Invariant Theory, and Spin.)

Above we have seen that $Ph(H)$ and therefore also $Ph(S(l))$ are moduli spaces of interest. We know that $U(1)$ acts on the components B_o and B_x, conferring a complex structure on the tangent bundle of $\underline{H}$. Moreover, the fundamental gauge group,

$$G := SU(2) \times SU(3),$$

acts on the complexified tangent bundle of $\underline{H}$, i.e. there is a principal G-bundle $\tilde{G}$ defined over $\underline{H}$, acting on $\Theta_{\underline{H}}$. Notice that if we choose a velocity $\mathbf{v}$, i.e. a directed line l in a tangent space of $\underline{H}$, and the corresponding Minkowski space-time defined by this directed tangent-line, then the action of G on the tangent space of this space-time, is trivially invariant under Lorentz boosts, since it, of course, leaves $H(l)$ fixed. This suggests that if we had a natural extension of the action of $\tilde{G}$ to the non-commutative algebra $Ph(H)$ then the non-commutative *orbit spaces*:

$$\underline{I} := Simp(Ph(H) : \tilde{G}),$$

or

$$\underline{I} := Simp(Ph(H) : \tilde{\mathfrak{g}}),$$

would have been a prime target for mathematical physics, see [18]. However, this seems not to be possible unless one gives up the commutativity of the base space H. We shall therefore, at this moment, restrict ourselves to a rather simple special case, which should be sufficiently general for our purpose.

Picking a line $l \subset \mathbf{E}^3$ and, as above, considering the subspace $\underline{S}(l)$, we see immediately that the trivial principal bundle, $SU(2)$, as well as the trivial $su(2)$-bundle (and therefore also the complexified $su(2)$, isomorphic to $sl_2(\mathbf{C})$), acts on the restriction of the tangent bundle Θ_H to $\underline{H}(l)$. $SU(2)$ and so also $sl_2(\mathbf{C})$ therefore also acts on the complexified tangent-bundle of $\underline{S}(l)$.

The tangent bundle of $\underline{H}$, restricted to $\underline{H} - \Delta$ decomposes as,

$$\mathbf{C}B_o \oplus \mathbf{C}B_x \oplus \mathbf{C}A_{o,x},$$

and on the exceptional fibers of $\underline{\tilde{H}}$, it decomposes as,

$$\mathbf{C}C_o \times \mathbf{C}A_o \times \mathbf{C}\tilde{\Delta}.$$

Restricted to $\underline{S}(l)$ these bundles are natural representation of $su(2)$ on $\mathbf{C}B_o \oplus \mathbf{C}B_x$, and of $su(3)$ on $\mathbf{C}\tilde{\Delta}$. But, of course, $su(2)$ acts trivially on $Ph(H(l))$. The above discussion concerning invariant spaces imply that we should be interested in $\underline{I} = Simp(Ph(S(l)) : su(2))$ or the representation theory of $Ph(S(l)) \times U(su(2))$, where $U(\mathfrak{g})$ is the universal enveloping algebra of the Lie algebra $\mathfrak{g}$.

As an example, pick $\alpha \in su(2) \subset End_{\mathbf{C}}(\mathbf{C}^2)$. Using the *parity*-operator P, see the Introduction, it acts naturally on $\mathbf{C}B_o$ and on $\mathbf{C}B_x$, as α and $-\alpha$, respectively. So it acts as,

$$\gamma_0(\alpha) = \begin{pmatrix} \alpha & 0 \\ 0 & -\alpha \end{pmatrix},$$

on the complex rank 4 vector bundle, $\mathbf{C}B_o \oplus \mathbf{C}B_x$, the sections of which are the *spinors* of the physicists. Composed with the parity operator P, $\gamma(\alpha)$ acts like,

$$\gamma(\alpha) = \begin{pmatrix} 0 & \alpha \\ -\alpha & 0 \end{pmatrix},$$

i.e. like Diracs representation.

Fix now a representation of $su(2)$, given in terms of the generators,

$$\alpha_1 = \begin{pmatrix} i & 0 \\ 0 & -i \end{pmatrix}, \alpha_2 = \begin{pmatrix} 0 & 1 \\ -1 & 0 \end{pmatrix}, \alpha_3 = \begin{pmatrix} 0 & i \\ i & 0 \end{pmatrix},$$

the analogues of the Pauli matrices, and put, $\mathbf{L} = (\alpha_1, \alpha_2, \alpha_3)$.

Notice that, picking the basis of $\mathbf{CB}_o \oplus \mathbf{CB}_x$ given by $e_1 = (1, 0, -1, 0), e_2 = (0, 1, 0, -1), e_4 = (1, 0, 1, 0), e_5 = (0, 1, 0, 1)$, then e_1, e_2 correspond to light velocities, and e_4, e_5 correspond to zero velocities. Let $\mathbf{S}_{o,x}$ be the subspace generated by the light-velocities e_1, e_2, and let $\mathbf{I}_{ox}$ be the subspace generated by the 0-velocities e_4, e_5, then the parity operator P will permute $\mathbf{S}_{o,x}$ and $\mathbf{I}_{ox}$.

Any *particle*, i.e. any simple representation of $Ph(S(l)) \rtimes U(su(2))$, is now canonically a $su(2)$-representation, as in classical quantum theory, where one imposes such an action, a spin structure, and in particular a Casimir element $S := L^2 \in su(2), L^2 = \sum \alpha_i^2$. Notice also that we have here a double situation, a spin structure for the action on the light-velocities, S_{ox}, and an iso-spin structure for the action on the 0-velocities, $\mathbf{I}_{ox}$. Let us explain this a little better.

The Lie algebra $su(3)$ has a 2-dimensional Cartan subalgebra $\mathfrak{h}$, and two copies of $su(2)$, $\mathfrak{g}_1$, and $\mathfrak{g}_2$. We may pick, in an essentially unique way, $\mathfrak{g}_1$, such that it leaves the dt_0-direction of the tangent space invariant, together with a non-zero element $s \in \mathfrak{h} \cap \mathfrak{g}_1 \subset \mathfrak{h}_1$, where the last Lie algebra is the 1-dimensional Cartan Lie-algebra of $\mathfrak{g}_1$. We may, with an obvious matrix notation, pick

$$s = \begin{pmatrix} 1/2 & 0 \\ 0 & -1/2 \end{pmatrix},$$

and assume that $\mathfrak{h}$ has a basis given by s and the element,

$$y = \begin{pmatrix} 1/3 & 0 & 0 \\ 0 & 1/3 & 0 \\ 0 & 0 & -2/3 \end{pmatrix}.$$

Physicists call s the *isospin*, and y the *hypercharge*. They are commonly denoted T_z and Y, and of course, are dependent upon the choice of the direction, l, and therefore of $\underline{H}(l)$. The common eigenvectors for s and y are called *quarks*. They come as *up*-quark, as *down*-quark, or as *strange*-quark. Obviously, there is room for far more strange and colorful occupants of the tangent bundle of $\underline{S}(l)$, by combining the many observables that we have at hand.

If we consider the moduli space $\mathbf{H} = \tilde{\underline{H}}/Z_2$, see the Introduction, then it is clear that the *gauge group*, Z_2, maps B_o isomorphically onto B_x and vice versa. Let $f_+ : B_o \oplus B_x \to B_o \oplus B_x$ and $f_- : B_o \oplus B_x \to B_o \oplus B_x$ be the corresponding projections. Obviously we have,

$$\{f_-, f_+\} = 1.$$

We therefore have the 4-dimensional situation described in *Grand Picture*, and we see that the generator $\Sigma \in Z_2$, not to be confused with the Parity operator P discussed above, reverses spin and isospin, but conserves the hypercharge. Notice that for a point in $\tilde{\underline{\Delta}}$ the decomposition of the tangent bundle of $\underline{S}(l)$ is different. There we have,

$$T_{\tilde{\underline{H}},o'} = C_{o'}' \oplus A_{o'} \oplus \tilde{\Delta},$$

where $C_{o'} \oplus A_{o'}$ are light-velocities, and here $su(2)$ act only upon the first factor. Thus here we have no isospin, and no quarks! To end this sketch, notice also that the metric g defined on $\underline{H}$, in Example (4.12), is very similar to the Schwartzchild metric of a black hole, with horizon equal to the exceptional fibre. The area of this *black hole horizon*, bounding nothing, as a function on $\underline{\Delta}$, is a candidate for mass-density of the Universe. See also that the symmetry, i.e. the gauge group, corresponding to points in $\underline{\Delta}$ is different from the gauge group outside $\underline{\Delta}$. This is analogous to deformation theory, where the automorphism group of an object sees a spontaneous reduction along a deformation outside the modular stratum. It therefore seems to me that, in this purely mathematical model, one might find a correlation between a notion of mass-distribution, and a kind of Higgs mechanism.

Notice also that, introducing a metric on $S(l)$ and picking the Killing form on $\mathfrak{g}$, the procedure of the section *Connections and the Generic Dynamical Law*, will lead to equations of motion of Dirac type and generalizations related to the theory of *weak forces*. See a forthcoming paper of Olav Gravir Imenes.

4.10 Cosmology, Big Bang and All That

In the paper [20], we discussed the possibility of including a cosmological model in our toy-model of Time-Space. The 1-dimensional model we presented there was created by the deformations of the trivial singularity, $O := k[x]/(x)^2$. Using elementary deformation theory for algebras, we obtained amusing results, depending upon some rather bold mathematical interpretations of the, more or less accepted, cosmological vernacular. Here we shall go one step further on, and show that our toy-model, i.e. the moduli space, $\mathbf{H}$, of two points in the Euclidean 3-space, or its étale covering, $\tilde{H}$, is *created* by the (non-commutative) deformations of the obvious singularity in 3-dimensions, $U := k < x_1, x_2, x_3 > /(x_1, x_2, x_3)^2$.

In fact, it is easy to see that the versal space $W(U)$, of the deformation functor of the k-algebra U, as embedded in 3-space, contains a flat component (a room in the (commutative) modular suite, see [22]) isomorphic to $\underline{H} - \underline{\Delta}$, and that the modular stratum (the inner room) is reduced to the base point. Notice that we are working with the non-commutative model of the 3-dimensional space. This is, of course, not visible in dimension 1, and therefore not highlighted in the above mentioned paper.

The tangent space of the versal base $\underline{H}(U)$, of the deformation functor of U, as an algebra, is given, see e.g. [14], by the cohomology,

$$T_{\underline{W}(U), *} = H^1(k, U; U) = Hom_F(J, U)/Der,$$

where $\pi : F \to U$, is a surjection of a (non-comutative) free k-algebra F onto U, $J = ker(\pi)$, and Der, the vector space of restrictions of derivations $\delta \in Der_k(F, U)$ to J. Pick $F = k < x_1, x_2, x_3 >$. One easily obtains a basis, $\{\alpha_k^{i,j}\}_{i,j,k}$ for $Hom_F(J, U)$, given in terms of the expression of the values of any F-linear maps $\alpha \in Hom_F(J, U)$,

$$\alpha(x_i x_j) = \sum_k \alpha_k^{i,j} \epsilon_k,$$

where ϵ_k is the class of x_k in U. This $Hom_F(J, U)$, is the tangent space of the mini-versal deformation space, $\underline{W}(U)$, of the imbedded singularity. The k-vector space of derivations, $Der_k(F, U)$, is of dimension 9, but the restrictions to $J = (x_i x_j)$ are given in terms of,

$$\delta(x_i x_j) = \delta(x_i)\epsilon_j + \epsilon_i \delta(x_j),$$

so that, with the notations above, $\delta_k^{i,j} = \delta^i$ if $k = j$, $\delta_k^{i,j} = \delta^j$ if $k = i$, and $\delta_k^{i,i} = 2\delta^i$ if $k = i$, and 0 otherwise. In particular Der is of dimension 3, determined by the values $\delta^i := \delta(x_i)$.

A point of $\underline{H}$ is an ordered pair (o, x) of two points $o = (\alpha_1^1, \alpha_2^1, \alpha_3^1)$, $x = (\alpha_1^2, \alpha_2^2, \alpha_3^2)$. Consider now the sub vector space $T(2)$ of $Hom_F(J, U)$, generated by the linear maps defined by,

$$\alpha(x_i x_j) = \alpha_i^1 \epsilon_j + \alpha_j^2 \epsilon_i.$$

The expressions above show that $Der \subset T(2)$. Moreover, the quotient space, i.e. the subspace of the tangent space $T_{\underline{H}(U), *}$, defined by $T(2)$, can be represented by the maps,

$$\alpha(x_i x_j) = (\alpha_j^2 - \alpha_j^1)\epsilon_i,$$

i.e. by the vectors ox in Euclidean 3-space. The Lie algebra of infinitesimal automorphisms of U, i.e. $Der_k(U, U)$ acts on the tangent space of $T(2)$, as follows. If $\delta \in Der_k(U)$, then for any $\alpha \in T(2)$, we have,

$$\delta(\alpha)(x_i, x_j) = -\sum_l \delta_i^l \alpha^1(x_l)\epsilon_j - \sum_l \delta_j^l \alpha^1(x_l)\epsilon_i,$$

inducing a distribution $\tilde{\Delta}$ of $\underline{H} - \underline{\Delta}$, and leaving only 0 invariant, in the tangent space of $\underline{W}(U)$ at the base point, $*$. This shows that the part of the modular substratum of $\underline{H}(U)$, sitting in $T(2)$ is reduced to $*$.

We may identify $*$, i.e. the singularity U, with the point $\alpha_i^p = 0, i = 1, 2, 3, p = 1, 2$, contained in $\underline{\Delta}$, but recall that the points of $\underline{\Delta}$ do not sit in $\tilde{\underline{H}}$. Moreover, the fact that the tangent space of $*$ does not contain any 0-velocity vectors, proves that $*$ is a fixed point of the versal space $\underline{W}$, and therefore of $\underline{\Delta}$.

This shows that $\underline{H} - \underline{\Delta}$ is a natural subspace of $\underline{W}(U)$, defined by the equations,

$$(x_i - \alpha_i^1)(x_j - \alpha_j^2) = 0, \; i, j = 1, 2, 3.$$

Notice that this system of equations is not equivalent to the commutative algebra equations,

$$(x_1 - \alpha_1^1, x_2 - \alpha_2^1, x_3 - \alpha_3^1)(x_1 - \alpha_1^2, x_2 - \alpha_2^2, x_3 - \alpha_3^2) = 0$$

which gives us the set of pairs of unordered points in 3-space, i.e. part of **H**.

Given any point $\underline{t} = (o, x) \in \tilde{\underline{H}}$, there is a translation $\omega(o, x) \in \tilde{\Delta}(\underline{t})$, the meaning of which should be clear, translating the vector $\bar{ox}$ so that its middle point coincides with $*$. The length, ω of $\omega(o, x)$ is our *Cosmological Time*.

For every point $(o, x) \in \tilde{\underline{H}}$ there is a vector $\xi \in T_{U,0} = \tilde{c}(*)$ such that $\bar{ox}$ is a translation by $\omega(o, x)$ of $\omega\xi$. We may express this by saying that (o, x) is created by the tangent vector ξ of U at the cosmological time ω.

Since $*$ is a fixed point in $\underline{\Delta} \simeq \mathbf{E}^3$, the geometry of our space $\tilde{\underline{H}}$ changes. The natural symmetry (gauge) group operating on $\mathbf{E}^3$ is now $G := Gl(3)$. The action of $\mathfrak{g} := lieG$ defines a distribution $\tilde{\mathfrak{g}}$, in $\tilde{\underline{H}}$, which is different from $\tilde{\Delta}$. In particular, the diagonal group, $K^* \subset G$, generates a 1-dimensional distribution, $\tilde{\omega} \subset \tilde{\mathfrak{g}}$, in $\tilde{\underline{H}}$.

Notice that the subspace $U(\omega) \subset \tilde{\underline{H}}$ with a fixed cosmological time ω, is of dimension 5. The tangent space of any point $\underline{t} \in U(\omega)$ has a 2-dimensional subspace, $\tilde{\beta}(\underline{t})$, normal to $\tilde{\omega}(\underline{t})$, which is not a subspace of $\tilde{\Delta}(\underline{t})$.

The metric g defined on $\tilde{\underline{H}}$ may now be written,

$$g = \lambda(\underline{t})d\omega^2 + g(\omega)$$

Here $g(\omega)$ is the metric induced on $U(\omega)$.

Notice that the maximal distribution $\tilde{\underline{c}}$, the *light-velocities* is no longer defined solely in terms of the metric g.

Now recall that time, in our model, is the metric g. Therefore we may look at the equations above as the following formula,

$$\lambda(\underline{t})d\omega^2 = dt^2 - g(\omega)$$

where t is time. Cosmological time is therefore of the same form as Einstein's *proper time*. This should be compared with the Einstein-de Sitter metric of the *elementary cosmological model*, see [27],

$$-g = du^4 \otimes du^4 - (R^2(u^4)) \sum_{i,j=1}^{3} du^\nu \otimes du^\nu,$$

and the Friedmann-Robertson-Walker metric, see [2],

$$-ds^2 = d\tau^2 = dt^2 - R^2(t)(\frac{dr^2}{(1 - kr^2)} + r^2 d\Omega^2).$$

It seems to us, that our model has an advantage over these, ad-hoc, models. Accepting Big Bang as the *creator* of space, via the (uni)versal family of the primordial U, changes the geometry of $\tilde{\underline{H}}$. In this picture, the point $(o, x) =: \underline{t} \in \tilde{\underline{H}}$ is created from $* = U$, via the tangent vector $v := 1/\omega \bar{o}x$ in the tangent space of $* = U$, in the direction, defined by *the middle point* of $(o, x = \underline{t})$, of $\underline{\Delta}$, and in the time-span defined by the cosmological time ω. This defines a half-line $*\underline{t}$, extending from $*$ through $\underline{t} \in \underline{H}$.

The visible universe ,$\mathbf{V}(o)$, for $ME=(o,o)$ should be the union of all world curves of *photons*, leaving $*$, reaching ME. This should coincide with the union of all *solutions* of the Lagrangian $\lambda(\underline{t})d\omega^2 = dt^2 - g(\omega)$, or with the corresponding geodesics of this non-positive definite metric. This implies that the universe must be curved, just like the picture drawn in [20], page 262, suggests. In that paper we just worked with a 1-dimensional real space. Never the less, the situation here, in our 3-dimensional picture, is basically the same, complete with a Hubble-constant, that is not really constant, etc.

If we want a catchy way to express these basic properties of our model, we might say that, U, the Big Bang, that created our (visible) world, is equivalent to any center of the exceptional fiber in $\tilde{\underline{H}}$, which, of course does not really exist, and so the *infinite small* has the same structure as the origin of the world, U.

In particular, see Example (4.14), we find that the symmetry brake of our model for quarks, in $S(l)$, where the $su(3)$ symmetry is broken by the unique 0-velocity dt_0, now could have been made global, using the cosmological distribution instead of dt_0.

Making all this fit with contemporary quantum theory and cosmology is, however, not an easy task. There are serious interpretational difficulties here, as well as in most papers we have seen, on cosmology. We shall therefore leave it for now, and hopefully be able to return to this later.

Chapter 5

Interaction and Non-commutative Algebraic Geometry

5.1 Interactions

Given a dynamical system σ of, order 2. A *particle*, $\tilde{V}$, that we know occured at some point $\underline{t} \in Simp_n(\mathbf{A}(\sigma))$, producing a simple representation $V := \tilde{V}(\underline{t})$ will after some time τ have develloped into the particle sitting at a point on the integral curve γ defined by the vectorfield ξ of σ, at a distance τ in $Simp_n(\mathbf{A}(\sigma))$ (we are of course assuming the field k is contained in the real numbers). Now, this may well be a point on the border of $Simp_n(\mathbf{A}(\sigma))$, i.e. in $\Gamma_n = Simp(C(n)) - U(n)$, where it *decays* into an indecomposable, or into a semi-simple, representation, i.e. into two or more new particles $\{V_i \in Simp_{n_i}(\mathbf{A}(\sigma)), n = \sum n_i\}$. What happens now is taken care of by the following scenario: If the different particles we have produced are not interacting, each of the new particles should be considered as an independent object, evolving according to the Dirac derivation δ. However, if the particles we have produced are interacting, we have a different situation. Notice first that for $n = 1$, we have a canonical morphism of schemes,

$$Simp_1(\mathbf{A}(\sigma)) \longrightarrow Simp_1(A)$$

and a canonical vector-field ξ in $Simp_1(\mathbf{A}(\sigma))$, the *phase space*. Given any point of $Simp_1(A)$, the *configuration space*, and any tangent-vector at this point, there is an integral curve of ξ in $Simp_1(\mathbf{A}(\sigma))$, through the corresponding point, projecting down to a fundamental curve in the configuration space.

For $n \geq 2$ the spaces $Simp_n(\mathbf{A}(\sigma))$ and $Simp_n(A)$ are, however, totally different and without any easy relations to each other.

Let now $v_i \in Simp_{n_i}(\mathbf{A}(\sigma))$, $i = 1, 2$ be two points of $Simp(\mathbf{A}(\sigma))$ corresponding to representations V_1, V_2, maybe in different components, and/or ranks. Consider their components, i.e. the universal families in which they are contained,

$$\tilde{\rho}_i : \mathbf{A}(\sigma) \longrightarrow End_{C(n_i)}(\tilde{V}_i)$$

The Dirac derivation, δ, defines derivations,

$$[\delta_i] : \mathbf{A}(\sigma) \longrightarrow End_{C(n_i)}(\tilde{V}_i)$$

and

therefore also the fundamental vector-fields, $\partial_i \in Ext^1_{\mathbf{A}(\sigma)\otimes_k C(n_i)}(\tilde{V}_i, \tilde{V}_i)$, and $\xi_i \in Der(C(n_i))$.

Definition 5.1.1. Let B be any finitely generated k-algebra. We shall say that the components, $C_1 \subseteq Simp_{n_1}(B)$, $C_2 \subseteq Simp_{n_2}(B)$, or the corresponding *particles* $\tilde{V}_i$, i=1,2, are *non-interacting* if

$$Ext^1_B(V_1, V_2) = 0, \forall v_1 \in C_1, \forall v_2 \in C_2,$$

where v_i is the isoclass of V_i Otherwise they *interact*.

Suppose now that the points v_1 and v_2, sit in $Simp_{n_1}(\mathbf{A}(\sigma))$ and $Simp_{n_2}(\mathbf{A}(\sigma))$, respectively. Physically, we shall consider this as an observation of two particles, $\tilde{V}_1$ and $\tilde{V}_2$ in the state V_1 and V_2, at some instant. If the two particles are non-interacting, the resulting entity, considered as the the sum $V := V_1 \oplus V_2$, of dimension $n := n_1 + n_2$, as module over $\mathbf{A}(\sigma)$, will stay, as time passes, a sum of two simples.

If V_1 and V_2 interact , this may change. To explain what may happen, we have to take into consideration the non-commutativity of the geometry of PhA. In particular, we have to consider the non-commutative deformation theory, see Chapter 3, and [16]. Consider the deformation functor,

$$Def_{\{V_1,V_2\}} : \underline{a}_2 \longrightarrow \underline{Sets},$$

or, if we want to deal with more points, say a finite family V_i, $i = 1, 2, ..., r$, the deformation functor,

$$Def_{\{V_i\}} : \underline{a}_r \longrightarrow \underline{Sets},$$

and its formal moduli,

$$H := \begin{pmatrix} H_{1,1} & ... & H_{1,r} \\ & ... & . \\ H_{r,1} & ... & H_{r,r} \end{pmatrix},$$

together with the versal family, i.e. the essentially unique homomorphism of k-algebras,

$$\tilde{\rho} : \mathbf{A}(\sigma) \longrightarrow \begin{pmatrix} H_{1,1} \otimes End_k(V_1) & \dots & H_{1,r} \otimes Hom_k(V_1, V_r) \\ \cdot & \dots & \cdot \\ H_{r,1} \otimes Hom_k(V_r, V_1) & \dots & H_{r,r} \otimes End_k(V_r) \end{pmatrix}.$$

This is, in an obvious sense, the *universal interaction*. However, we need a way of specifying which interactions we want to consider. This is the purpose of the following, tentative, definition,

Definition 5.1.2. Given $v_i \in Simp(A(\sigma))$, $i = 1, ..., r$. An *interaction mode* for the corresponding family of modules $\{V_i\}$, i=1,...,r, is a right $H(\{V_i\})$-module M.

An interaction mode is a kind of *higher order preparation*, see Chapter 2. It consists of a rule, telling us, for the given family of r points, $v_i \in Simp_{n_i}(A(\sigma))$ how to *prepare* their *interactions*. The structure morphism $\phi : H(\{V_i\}) \to End_k(M)$, fixes all relevant *higher order* momenta, i.e. ϕ evaluates all the tangents between these modules, and by the *Beilinson-type* Theorem, see Chapter 3, creates a new $A(\sigma)$-module.

In fact, an interaction mode induces a homomorphism,

$$\kappa(M) : A(\sigma) \longrightarrow End_k(\tilde{V}),$$

where $\tilde{V} := \oplus_{i=1,...,r} M_i \otimes V_i$.

Thus, we have constructed a new n-dimensional $A(\sigma)$-module, which may be decomposable, indecomposable or simple, depending on the interaction mode we choose, and, of course, depending upon the tangent structure of the moduli space $Simp(A(\sigma))$.

Assuming the impossible, that our k-algebra of observables, $\mathbf{A}(\sigma)$, consisted of all observables that our curiosity fancied. Assume moreover that the notion of interaction mode, i.e. a right $H(Simp(\mathbf{A}(\sigma)))$-module, made sense, then we might talk about a QFT-theory of the UNIVERSE. The piece-vise defined curves, $\gamma_0, \gamma_1, ..., \gamma_l, ..$, corresponding to successive choices at each *moment* of decay, of a new interaction mode, and therefore of a new integral curve in the relevant time-space $Simp_N(\mathbf{A}(\sigma))$, might be called a *history*.

Determining the conditions for de-coherence, and the assignment of probabilities for these choices, will be left for now. See [6], and [7]. The problem of time, in this context, observed from inside or outside of the universe, will also be postponed, maybe á la calendes grecques.

5.2   Examples and Some Ideas

Example 5.1. Let us consider the case of two point-objects interacting in 3-space. Go back to the Example (2.1), (iii), and let us look at two fields,

$$\phi(i) : A_0 := k[x_1, x_2, x_3] \to k[\tau], \ i = 1, 2,$$

inducing,

$$Ph(\phi(i)) : A := Ph(k[x_1, x_2, x_3]) \to Ph(k[\tau]), \ i = 1, 2.$$

This corresponds to two curves $\gamma(i)$, $i = 1, 2$, in the phase space, $\underline{A}$, parametrized by the same clock-time τ. Start with $\tau = 0$, and let this time correspond to the two points,

$$\phi(i, 0) = v_i := (q_i, p_i) \in \underline{A}, \ , \ i = 1, 2.$$

Let $V_i := k(v_i)$, $i = 1, 2$, assume $q_1 \neq q_2$, and use (2.1)(iii), to find that the formal moduli of the family $\{V_1, V_2\}$ of A-modules, has the form,

$$\begin{pmatrix} \hat{A}_{p_1} & < \tau_{1,2} > \\ < \tau_{2,1} > & \hat{A}_{p_2} \end{pmatrix}$$

where $\tau_{i,j}$ is a generator for $Ext^1_A(V_i, V_j)^* \simeq k$.

Since $Hom_k(V_i, V_j) = k$, $i, j = 1, 2$, we have a natural, versal family, i.e. a morphism,

$$A \to \begin{pmatrix} \hat{A}_{p_1} & < \tau_{1,2} > \\ < \tau_{2,1} > & \hat{A}_{p_2} \end{pmatrix},$$

An interaction mode between the two point-particles, defined by $\phi(i)$, $i = 1, 2$, is now given by evaluating the $\tau_{i,j}$, and expressing the two fields ϕ_i by the corresponding morphism,

$$\phi : A \to \begin{pmatrix} Ph(k[\tau]) & 0 \\ 0 & Ph(k[\tau]) \end{pmatrix}.$$

A force law should now be given by some elements,

$$\psi_{i,j}(\tau) \in Ext^1_A(\phi_i(\tau), \phi_j(\tau)).$$

by, say, putting,

$$\psi_{i,j}(\tau) = \phi(i, j : \tau) \cdot \xi_{i,j},$$

where $\xi_{i,j}$ is the generator of $Ext^1_A(\phi_i(\tau), \phi_j(\tau))$ found in (2.1), (iii).

In the general case, we need a notion of *hypermetric* to formulate a reasonable theory of interaction. This will, I hope, be the topic of a forthcoming paper.

Example 5.2. Let us consider the notion of interaction between two *particles*, $V_i := k(v_i) \in k[x_1, ..., x_r]$, $i = 1, 2$, in the above sense. Look at the $A_0 := k[x_1, ..., x_r]$-module $V := V_1 \oplus V_2$, i.e. the homomorphism of k-algebras, $\rho_0 : A_0 \to End_k(V)$, and let us try to extend this module-structure to a representation,

$$\rho : Ph^{\infty} A_0 \to End_k(V).$$

We have the following relations in $Ph^{\infty} A_0$:

$$[x_i, x_j] = 0$$

$$[dx_i, x_j] + [x_i, dx_j] = 0$$

$$....$$

$$\sum_{l=0}^{p} \binom{p}{l} [d^l t_i, d^{p-l} t_j] = 0.$$

Put,

$$\rho_0(x_i) = \rho_0(d^0 x_i) = \begin{pmatrix} x_i(1) & 0 \\ 0 & x_i(2) \end{pmatrix} =: \begin{pmatrix} \alpha_i^0(1) & 0 \\ 0 & \alpha_i^0(2) \end{pmatrix},$$

and, $\alpha_i^0(r, s) := x_i(r) - x_i(s)$, $r, s = 1, 2$. Let, for $q \geq 0$,

$$\rho(d^q x_i) = \begin{pmatrix} \alpha_i^q(1) & r_i^q(1, 2) \\ r_i^q(2, 1) & \alpha_i^q(2) \end{pmatrix},$$

Now, compute, for any $p \geq k$,

$$[\rho(d^k x_i), \rho(d^{p-k} x_j)] =$$

$$\begin{pmatrix} r_i^k(1, 2) r_j^{p-k}(2, 1) - r_j^{p-k}(1, 2) r_i^k(2, 1) & r_j^{p-k}(1, 2) \alpha_i^k(1, 2) + r_i^k(1, 2) \alpha_j^{p-k}(2, 1) \\ r_i^k(2, 1) \alpha_j^{p-k}(1, 2) + r_j^{p-k}(2, 1) \alpha_i^k(2, 1) & r_i^k(2, 1) r_j^{p-k}(1, 2) - r_j^{p-k}(2, 1) r_i^k(1, 2) \end{pmatrix}$$

and observe that,

$$[\rho(d^p x_i), \rho(x_j)] + [\rho(x_i), \rho(d^p x_j)] =$$

$$\begin{pmatrix} 0 & r_i^p(1, 2) \alpha_j^0(2, 1) + r_j^p(1, 2) \alpha_i^0(1, 2) \\ r_i^p(2, 1) \alpha_j^0(1, 2) + r_j^p(2, 1) \alpha_i^0(2, 1) & 0 \end{pmatrix}$$

After some computation we find the following condition for these matrices to define a homomorphism ρ, *independent of the choice of diagonal forms*,

$$r_i^k(r, s) = \sum_{l=0}^{k} \binom{k}{l} \sigma_{k-l} \alpha_i^l(r, s); \ r, s = 1, 2,$$

where the sequence $\{\sigma_l\}$, $l = 0, 1, ...$ is an arbitrary sequence of *coupling constants*, with $\sigma_0 = 0$ and σ_l of *order l*. By recursion, we prove that this is true, for $k \leq p - 1$, therefore $r_i^k(1,2) = -r_i^k(2,1)$, and so the diagonal elements above vanish, i.e.

$$r_i^k(1,2)r_j^{p-k}(2,1) - r_j^{p-k}(1,2)r_i^k(2,1) = 0.$$

The general relation is therefore proved if we can show that with the above choice of $r_i^k(r,s)$ we obtain, for every $p \geq 0$,

$$\sum_{k=0}^{p} \binom{p}{k}(r_j^{p-k}(1,2)\alpha_i^k(1,2) + r_i^k(1,2)\alpha_j^{p-k}(2,1)) = 0,$$

and this is the formula,

$$\sum_{k=0}^{p} \binom{p}{k} \sum_{l=0}^{p-k} \binom{p-k}{k} \sigma_{p-k-l}\alpha_j^l(1,2)\alpha_i^k(1,2)+$$

$$\sum_{k=0}^{p} \binom{p}{k} \sum_{l=0}^{k} \binom{k}{l} \sigma_{k-l}\alpha_i^l(1,2)\alpha_j^{p-k}(2,1) =$$

$$\sum_{k=0}^{p} \binom{p}{k} \sum_{l=0}^{k} \binom{k}{l} \sigma_{k-l}(\alpha_j^l(1,2)\alpha_i^{p-k}(2,1) - \alpha_i^l(1,2)\alpha_j^{p-k}(2,1)) = 0,$$

Notice that the relations above are of the same form for any commutative coefficient ring C, i.e. they will define a homomorphism,

$$\rho : Ph^\infty A_0 \to M_2(C),$$

for any commutative k-algebra C. Now, consider the Dirac time development $D(\tau) = exp(\tau\delta)$ in $\mathbf{D}$, the completion of $Ph^\infty A_0$. Composing with the morphism ρ defined above, we find a homomorphism,

$$\rho(\tau) : Ph^\infty(A_0) \to M_2(k[[\tau]]),$$

where,

$$X_i := \rho(\tau)(t_i) = \begin{pmatrix} \Phi_i(1) & \Phi_i(1,2) \\ \Phi_i(2,1) & \Phi_i(2), \end{pmatrix}$$

and

$$\Phi_i(r) = \sum_{n=0}^{\infty} 1/(n!)\tau^n \cdot \alpha_i^n(r), \ r = 1,2,$$

$$\Phi_i^0(r,s) = \sum_{n=0}^{\infty} 1/(n!)\tau^n \cdot \alpha_i^n(r,s), \ r,s = 1,2,$$

$$\sigma = \sum_{n=0}^{\infty} 1/(n!)\tau^n \cdot \sigma_n,$$

$$\Phi_i(r,s) = \sigma \cdot \Phi_i^0(r,s), \ r,s = 1,2.$$

This must be the most general, Heisenberg model, of motion of our two particles, clocked by τ. Observe that the *interaction acceleration* $\Phi_i(r, s)$ is *pointed from r to s*, just like in physics! The formula above, is now seen to be a consequence of the obvious equality of the two products of the formal power series, $\sigma \cdot (\Phi_i^0(1,2)\Phi_j^0(1,2))$ and $\sigma \cdot (\Phi_j^0(1,2)\Phi_i^0(1,2))$, just compare the coefficients of the resulting power series. What we have got is nothing but a formula for *commuting matrices* $\{X_i\}_{i=1}^d$ in $M_2(k[[\tau]])$, since for such matrices we must have,

$$\frac{d^n}{d\tau^n}[X_i, X_j] = 0, \ n \geq 1.$$

The eigenvalues $\lambda_i(1,\tau)$, and $\lambda_i(2,\tau))$ of X_i describes points in the space that should be considered the *trajectories* of the two points under interaction. This is OK, at least as long as we are able to label them by 1 and 2 in a continuous way with respect to the clock time τ. If all *coupling constants*, σ_n, $n \geq 0$, vanish, then the system is simply given by the two curves $\Phi(r) := (\Phi_1(r), \Phi_2(r), ..., \Phi_d(r))$, $r = 1, 2$, where d is the dimension of A_0. In general, the eigenvalues of X_i are given by,

$$\lambda_i(r) = 1/2 \cdot (\Phi_i(1) + \Phi_i(2))$$
$$(-1)^r 1/2 \sqrt{(\Phi_i(1) + \Phi_2(2))^2 - 4(\Phi_i(1) \cdot \Phi_i(2) + \sigma^2(\Phi_i(1) - \Phi_i(2))^2)},$$

for $r = 1, 2$. Clearly,

$$1/2(\lambda_i(1) + \lambda_i(2)) = 1/2(\Phi_i(1) + \Phi_i(2))$$
$$(\lambda_i(1) - \lambda_i(2))^2 = (\Phi_i(1) + \Phi_2(2))^2 - 4(\Phi_i(1) \cdot \Phi_i(2) + \sigma^2(\Phi_i(1) - \Phi_i(2))^2)$$
$$= (1 - 4\sigma^2)(\Phi_i(1) - \Phi_i(2))^2.$$

Denote by, $\lambda(r) = (\lambda_1(r), \lambda_2(r), ..., \lambda_d(r))$, $r = 1, 2$, the vectors corresponding to the eigenvalues, and by,

$$o := 1/2(\lambda(1) + \lambda(2)) = 1/2(\Phi(1) + \Phi(2))$$

the common median, and put,

$$R_0 := |(\Phi(1) - \Phi(2)|$$
$$R := |\lambda(1) - \lambda(2)|,$$

then

$$R = \sqrt{(1 - 4\sigma^2)}R_0.$$

Choose coupling constants such that,

$$\frac{d^2}{d\tau^2}R = rR^{-2},$$

where r is a constant we should expect Newton like interaction. If $\sigma \geq 1/2$, the point particles are confounded, the eigenvalues of X_i become imaginary, and the result is no longer obvious. If we pick $\sigma = \sqrt{1 - r^2 R_0^{-2}}$, the relative motion will be circular, with constant radius r about o.

Example 5.3. Let B be the free k-algebra on two non-commuting symbols, $B = k < x_1, x_2 >$, and see Example (2.14). Let P_1 and P_2 be two different points in the (x_1, x_2)-plane, and let the corresponding simple B-modules be V_1, V_2. Then, $Ext_B^1(V_1, V_2) = k$. Let Γ be the quiver,

$$V_1 \longleftrightarrow V_2,$$

then an interaction mode is given by the following elements: First the formal moduli of $\{V_1, V_2\}$,

$$H := \begin{pmatrix} k < u_1, u_2 > & < t_{1,2} > \\ < t_{2,1} > & k < v_1, v_2 > \end{pmatrix},$$

then the k-algebra,

$$k\Gamma := \begin{pmatrix} k & k \\ 0 & k \end{pmatrix},$$

and finally a homomorphism,

$$\phi : H \longrightarrow k\Gamma$$

specifying the value of $\phi(t_{1,2}) \in Ext_B^1(V_1, V_2)$. Since $Hom_k(V_i, V_j) = k$, we obtain $V = k^2$, and we may choose a representation of $\phi(t_{1,2})$ as a derivation, $\psi_{1,2} \in Der_k(B, Hom_k(V_1, V_2))$, such that the B-module $V = k^2$ is defined by the actions of x_1, x_2, given by,

$$X_1 := \begin{pmatrix} \alpha_1 & 1 \\ 0 & \alpha_2 \end{pmatrix}, \quad X_2 := \begin{pmatrix} \beta_1 & 0 \\ 0 & \beta_2 \end{pmatrix},$$

where $P_1 = (\alpha_1, \beta_1)$ and $P_2 = (\alpha_2, \beta_2)$. V is therefore an indecomposable B-module, but not simple. If we had chosen the following quiver,

$$V_1 \longleftrightarrow^{\epsilon_{1,2}}_{\epsilon_{2,1}} V_2,$$

where $\epsilon_{i,j}\epsilon_{j,i} = 0$, $i, j = 1, 2$, then the resulting B-module $V = k^2$ would have been simple, represented by,

$$X_1 := \begin{pmatrix} \alpha_1 & 1 \\ 0 & \alpha_2 \end{pmatrix}, \quad X_2 := \begin{pmatrix} \beta_1 & 0 \\ 1 & \beta_2 \end{pmatrix}.$$

In general, if $B = \mathbf{A}(\sigma)$, where (σ) is a dynamical system with Dirac derivation δ, any interaction mode producing a simple module V, thus a point $v \in Simp(\mathbf{A}(\sigma))$, represents a *creation* of a new particle from the information contained in the interacting constituencies. Moreover, any family of state-vectors $\psi_i \in V_i$, produces a corresponding state -vector $\psi := \sum_{i=1,\ldots} \psi_i \in V$, and Theorem (3.3) then tells us how the evolution operator acts on this new state-vector. If the created new particle V is not simple, the Dirac derivation $\delta \in Der_k(\mathbf{A}(\sigma))$, will induce a tangent vector $[\delta](V) \in Ext^1_{PhA}(V, V)$ which may or may not be modular, or pro-representable, which means that the *particles* V_i, when integrated in this direction, may or may not continue to exist as distinct particles, with a non-trivial endomorphism ring, or, with a Lie algebra of automorphisms, equal to k^2. If they do, this situation is analogous to the case which in physics is referred to as the *super-selection rule*. Or, if $[\delta] \in Ext^1_{\mathbf{A}(\sigma)}(V, V)$ does not sit (or stay) in the modular stratum, the particle V looses auto-morphismes, and may become indecomposable, or simple, instantaneously. We may thus create new particles, and we have in Example (4.7) discussed the notion of lifetime for a given particle. In particular we found that the harmonic oscillator had ever-lasting particles of k-rank 2. If, however, we forget about the *dynamical system*, and adopt the more *physical* point of view, picking a *Lagrangian*, and its corresponding *action*, we may easily produce particles of finite lifetime.

Example 5.4. Let, as in (4.6) $A := PhA_0 = k < x, dx >$, with $A_0 = k[x]$ and put $x =: x_1$, $dx =: x_2$. Consider the curve of two-dimensional simple A-modules,

$$X_1 = \begin{pmatrix} 0 & 1+t \\ 0 & t \end{pmatrix}$$

$$X_2 = \begin{pmatrix} t & 0 \\ 1+t & 0 \end{pmatrix},$$

either as a free particle, with Lagrangian $1/2dx^2$, or as a harmonic oscillator with Lagrangian $1/2dx^2 + 1/2x^2$. The action is, in the first case, $S = 1/2TrX_2^2 = t^2$, and in the second case, $S = 1/2Tr(X_2^2 + 1/2X_1^2) = 2t^2$. Thus the Dirac-derivation becomes $\nabla S = t\frac{\partial}{\partial t}$, or $\nabla S = 2t\frac{\partial}{\partial t}$. Computing the Formanek center f, see (3.6), we find,

$$f(t) = t^2(1+t)^2 - (1+t)^4.$$

The corresponding particle, born at $t < 0$, decays at $t = -1/2$, and thus has a finite lifetime. Of course, the parameter t in this example, is not our *time*, and the curve it traces is not an integral curve of the dynamic system of the harmonic oscillator, see (3.7). This shows that one has to be careful about mixing the notions of dynamic system, and the dynamics stemming from a Lagrangian-, or from a related action-principle.

Example 5.5. Suppose we are given an element $v \in Simp(A(\sigma))$, and consider the monodromy homomorphism,

$$\mu(v) : \pi_1(v; Simp_n(A(\sigma))) \rightarrow Gl_n(k).$$

If v is Fermionic, then there exist a loop in $Simp_n(A(\sigma))$ for which the monodromy is non-trivial. Assume the tangent of this loop at v is given by $\xi \in Ext^1_{A(\sigma)}(V, V)$. Since this tangent has no *obstructions* it is reasonable to assume that there is a quotient,

$$H(\{V, V\}) \rightarrow \begin{pmatrix} k & f^- \\ f^+ & k \end{pmatrix},$$

with $f^- f^+ = f^+ f^- = 1$. This would give us an $A(\sigma)$-module, with structure map,

$$A(\sigma) \rightarrow \begin{pmatrix} End_k(V) & f^- \otimes_k End_k(V) \\ f^+ \otimes_k End_k(V) & End_k(V) \end{pmatrix},$$

i.e. a simple $A(\sigma)$-module of Fermionic type, see Chapter 4, Grand Picture etc. Notice that, for a given connection on the vector-bundle $\tilde{V}$, the correct monodromy group to consider for the sake of defining Bosons and Fermions, etc., should probably be the infinitesimal monodromy group generated by the derivations of the curvature tensor, R.

In physics, interactions are often represented by tensor products of the representations involved. For this to fit into the philosophy we have followed here, we must give reasons for why these tensor products pop up, seen from our moduli point of view. It seems to me that the most natural point of view might be the following: Suppose A is the moduli algebra parametrizing some objects $\{X\}$, and B is the moduli for some objects $\{Y\}$, then considering the *product*, or rather, the pair, (X, Y), one would like to find the moduli space of these *pairs*. A good guess would be that $A \otimes_k B$ would be such a space, since it algebraically defines the product of the two moduli spaces. However, this is, as we know, too simplistic. There are no reasons why the pair of two objects, should deform independently, unless we assume that they do not fit into any *ambiant space*, i.e. unless the two objects are considered to sit in totally separate universes, and then we have done nothing, but doubling our model in a trivial way. In fact, we should, for the purpose of explaining the role of the *product*, assume that our entire universe is parametrized by the moduli algebra A, and accepting, for two objects X and Y in this universe, that the superposition, or the pair, (X, Y), correspond to a collection of, maybe new, objects parametrized by A. This is basically what we do, when we assume that X and Y are of the same sort, represented by modules V and W of some moduli k-algebra A, and then consider some tensor product of the representations, say $V \otimes_k W$, as a new representation, modeling a collection of new particles. As above we observe that the obvious moduli space of tensor-products of representations, or of the pairs (V, W) is $A \otimes A$. But since these representations should be of the same nature as any representation of A, this would, by universality, lead to a homomorphism of moduli algebras,

$$\Delta : A \to A \otimes A,$$

i.e. to a bialgebra structure on the moduli algebra A. This is just one of the reasons why mathematical physicists are interested in Tannaka Categories, and in the vast theory of quantum groups. For an elementary introduction, and a good bibliography, see [10]. See now that, if A_0 is commutative, and if we put $A = Ph(A_0)$, then there exist a canonical homomorphism,

$$\Delta : A \to A \otimes_{A_0} A.$$

In fact, the canonical homomorphism $i : A_0 \to Ph(A_0)$ identifies A_0 with the a sub-algebra of $A \otimes_{A_0} A$. Moreover, $d \otimes 1 + 1 \otimes d$ is a natural derivation, $A_0 \to A \otimes_{A_0} A.$, so by universality, Δ is defined. Thus, for representations of $A := Ph(A_0)$ there is a natural tensor product, $- \otimes_{A_0} -$. Thus, in (3.18), the tensor product of the fiber bundles defined on $\underline{S}(l)$,

$$PN := \tilde{\Delta} \otimes_H \tilde{\Delta} \otimes_H \tilde{\Delta},$$

is, in a natural way, a new representation of $Ph(S(l))$, the fibers of which is the triple tensor product,

$$(3) \otimes_k (3) \otimes_k (3)$$

of the Lie-algebra $su(3)$. The representation PN therefore splits up in the well-known swarm of *elementary particles*, among which, the proton and the neutron, see (4.18). Notice, finally, that the purpose of my notion of *swarm*, see [18], is to be able to handle a more complicated situation than the one above. One should be prepared to sort out the swarm of those representations of the known observables, that one would like to accept as models for physical objects, and then compute the parameter algebra best fitting this swarm. This is, in my opinion, one of the main objectives of a fully developed future non-commutative algebraic geometry.

Bibliography

[1] M. Artin (1969) *On Azumaya Algebras and Finite Dimensional Representations of Rings*, Journal of Algebra 11, 532-563 (1969).

[2] E. Elbaz (1995) *Quantique* ellipses/edition marketing S.A. (1995).

[3] L. Faddeev (1999) *Elementary Introduction to Quantum Field Theory. Quantum Fields and Strings* A Course for Mathematicians. Volume 1. American Mathematical Society. Institute for Advanced Study (1999).

[4] E. Formanek (1990) *The polynomial identities and invariants of $n \times n$ Matrices* Regional Conferenc Series in Mathematics, Number 78. Published for the Conference Board of the Mathematical Sciences by the American Mathematical Society, Providence, Rhode Island (1990).

[5] E. Fraenkel (1999) *Vertex algebras and algebraic curves* Séminaire Bourbaki, 52ème année, 1999-2000, no. 875.

[6] M. Gell-Mann (1994) *The Quark and the Jaguar*. Little, Brown and Company (1994).

[7] M. Gell-Mann and J. B. Hartle (1996) *Equivalent Sets of Histories and Multiple Quasiclassical Realms* arXiv:gr-qc/9404013v3, (5 May 1996).

[8] O. Gravir Imenes (2005) *Electromagnetism in a relativistic quantummechanic model* Master thesis, Matematisk institutt, University of Oslo, June 2005.

[9] S. Jøndrup, O. A. Laudal, A. B. Sletsjøe, *Noncommutative Plane Curves*, Forthcoming.

[10] C. Kassel (1995) *Quantum Groups*, Springer Graduate Texts in Mathematics 155 (1995).

[11] Etienne Klein and Michel Spiro, *Le Temps et sa Fleche*, Champs, Flammarion 2.ed. (1996).

[12] D. Laksov and A. Thorup (1999) *These are the differentials of order n*, Trans. Amer. Math. Soc. 351 (1999) 1293-1353.

[13] O. A. Laudal (1965) *Sur la théorie des limites projectives et inductives* Annales Sci. de l'Ecole Normale Sup. 82 (1965) pp. 241-296.

[14] O. A. Laudal (1979) *Formal moduli of algebraic structures*, Lecture Notes in Math.754, Springer Verlag, 1979.

[15] O. A. Laudal (1986) *Matric Massey products and formal moduli I* in (Roos,

J.E. ed.) Algebra, Algebraic Topology and their interactions Lecture Notes in Mathematics, Springer Verlag, vol 1183, (1986) pp. 218–240.

[16] O. A. Laudal (2002) *Noncommutative deformations of modules*, Special Issue in Honor of Jan-Erik Roos, Homology, Homotopy, and Applications, Ed. Hvedri Inassaridze. International Press, (2002). See also: Homology, Homotopy, Appl. 4 (2002) pp. 357-396.

[17] O. A. Laudal (2000) *Noncommutative Algebraic Geometry*, Max-Planck-Institut für Mathematik, Bonn, 2000 (115).

[18] O. A. Laudal (2001) *Noncommutative algebraic geometry*, Proceedings of the International Conference in honor of Prof. Jos Luis Vicente Cordoba, Sevilla 2001. Revista Matematica Iberoamericana.19 (2003) 1-72.

[19] O. A. Laudal (2003) *The structure of $Simp_n(A)$* (Preprint, Institut Mittag-Leffler, 2003-04.) Proceedings of NATO Advanced Research Workshop, Computational Commutative and Non-Commutative Algebraic Geometry. Chisinau, Moldova, June 2004.

[20] O. A. Laudal (2005) *Time-space and Space-times*, Conference on Noncommutative Geometry and Representatioon Theory in Mathematical Physics. Karlstad, 5-10 July 2004. Ed. Jürgen Fuchs, et al. American Mathematical Society, Contemporary Mathematics, Vol. 391, 2005.

[21] O. A. Laudal (2007) *Phase Spaces and Deformation Theory*, Preprint, Institut Mittag-Leffler, 2006-07. See also the part of the paper published in: Acta Applicanda Mathematicae, 25 January 2008.

[22] O. A. Laudal and G. Pfister (1988) *Local moduli and singularities*, Lecture Notes in Mathematics, Springer Verlag, vol. 1310, (1988).

[23] F.Mandl and G. Shaw (1984) *Quantum field theory*, A Wiley-Interscience publication. John Wiley and Sons Ltd. (1984).

[24] C. Procesi (1967) *Non-commutative affine rings*, Atti Accad. Naz. Lincei Rend.Cl. Sci. Fis. Mat. Natur. (8)(1967) 239-255.

[25] C. Procesi (1973): Rings with polynomial identities. Marcel Dekker, Inc. New York, (1973).

[26] I. Reiten(1985) *An introduction to representation theory of Artin algebras*, Bull.London Math. Soc. 17, (1985).

[27] R. K. Sachs and H. Wu (1977) *General Relativity for Mathematicians*, Springer Verlag, (1977).

[28] M. Schlessinger (1968) *Functors of Artin rings*, Trans. Amer. Math. Soc. vol.1. 30 (1968) 208-222.

[29] T. Schücker (2002) *Forces from Connes' geometry*, arXiv:hep-th/0111236v2, 7 June 2002.

[30] S. Weinberg (1995) *The Quantum Theory of Fields. Vol I, II, III* Cambridge University Press (1995).

Useful readings

[31] St. Augustin, *Les Confessions de Saint Augustin*, par Paul Janet. Charpentier, Libraire-Éditeur, Paris (1861).

[32] F. A. Berezin, *General Concept of Quantization*, Comm. Math. Phys. 40 (1975) 153-174.

[33] H. Bjar and O. A. Laudal, *Deformation of Lie algebras and Lie algebras of deformations*, Compositio Math. vol 75 (1990) pp. 69–111.

[34] Abraham Pais, *Niels Bohr's Times in physics, philosophy and polity*, Clarendon Press, Oxford (1991).

[35] T. Bridgeland-A. King-M. Reid, *Mukai implies McKay: the McKay correspondence as an equivalence of derived categories*, arXiv:math.AG/9908027 v2 2 May 2000.

[36] Marcus Chown, *The fifth element*, New Scientist, Vol 162 No 2180, pp. 28-32.(3 April 1999).

[37] Ali H. Chamseddine and Alain Connes, *The Spectral Action Principle*, Comm. Math. Phys. 186 (1997).

[38] Alain Connes, *Noncommutative Differential Geometry and the Structure of Space-Time*, The Geometric Universe. Science, Geometry, and the Work of Roger Penrose. Ed. by S.A. Huggett, L.J. Mason, K.P. Tod, S.T. Tsou, and N.M.J. Woodhouse. Oxford University Press (1998).

[39] André Comte-Sponville, *Pensées sur le temps*, Carnets de Philosophie, Albin Michel, (1999).

[40] Peter Coveney and Roger Highfield, *The Arrow of Time*, A Fawcett Columbine Book, Ballantine Books (1992).

[41] John Earman, Clark Glymour, and John Stachel *Foundations of Space-Time Theories*. University of Minnesota Press, Minneapolis, Vol VIII (1977).

[42] Ivar Ekeland, *Le meilleur des mondes possibles*, Editions du Seuil/science ouverte. (2000).

[43] S. A. Fulling, *Aspects of Quantum Field Theory in Curved Space-Time*, London Mathematical Society. Student Texts 17. (1996).

[44] G. W. Leibniz, *Nouveaux Essais IV, 16.*

[45] Misner, Thorne and Wheeler, *Gravitation*.

[46] David Mumford, *Algebraic Geometry I. Complex Projective Varieties*, Grundlehren der math. Wissenschaften 221, Springer Verlag (1976).

[47] Paul Arthur Schilpp, *Albert Einstein: Philosopher-Scientist*, The Library of Living Philosophers, La Salle, Ill. Open Court Publishing Co. (1970).

[48] C. T. Simpson, *Products of matrices. Differential Geometry, Global Analysis and Topology*, Canadian Math. Soc. Conf. Proc. 12, Amer.Math. Soc. Providence (1992) pp. 157-185.

[49] A. Siqveland, *The moduli of endomorphisms of 3-dimensional vector spaces.* Manuscript. Institute of Mathematics, University of Oslo (2001)

[50] Lee Smolin, *Rien ne va plus en physique!* Quai des siences. Dunod Paris (2007).

[51] P. Woit, *Not even wrong*, Vintage Books, London (2007).

Index